T0301366

Politics of Renewable Energy in China

NEW HORIZONS IN ENVIRONMENTAL POLITICS

Series Editor: Arthur Mol, *Chair and Professor in Environmental Policy, Director, Wageningen School of Social Sciences, Wageningen University, the Netherlands, Professor in Environmental Policy, Renmin University, Beijing*

The New Horizons in Environmental Politics series provides a platform for in-depth critical assessments of how we understand the many changes in the politics of nature, the environment and natural resources that have occurred over the last 50 years. Books in the series question how the environment is (re)defined, debated and protected; explore differences between countries and regions in environmental politics; analyse how actors do and do not collaborate around environment and natural resource conflicts; describe who wins and who loses and in what ways; and detail how to better study, analyse and theorize such developments and outcomes.

The series is designed to promote innovative cross-disciplinary analysis of the contemporary issues and debates influencing the various dimensions of environmental politics. Covering a diverse range of topics, the series will examine the political, economic and ethical aspects of environmental policy, governance and regulation. It brings together cutting edge research on environmental politics worldwide in order to shed light on, and explain current trends and developments.

With oversight from the Series Editor, Professor Arthur Mol – a noted specialist in the field of environmental politics at Wageningen University, the Netherlands – the New Horizons in Environmental Politics series comprises carefully commissioned projects from experts in the field including both academics and professionals. The audience for the series is global, and books in the series are essential reading for students, academics and professionals – in short, anyone with an interest in understanding the vital issues affecting environmental politics in the twenty-first century.

Titles in the series include:

American Environmental Policy
The Failures of Compliance, Abatement and Mitigation
Daniel Press

Environmental Governance through Partnerships
A Discourse Theoretical Study
Ayşem Mert

The Domestic Politics of Global Climate Change
Key Actors in International Climate Cooperation
Edited by Guri Bang, Steinar Andresen and Arild Underdal

Linking EU Climate and Energy Policies
Decision-making, Implementation and Reform
Jon Birger Skjærseth, Lars H. Gulbrandsen, Per Ove Eikeland and Torbjørg Jevnaker

The Challenges of Collaboration in Environmental Governance
Barriers and Responses
Edited by Richard D. Margerum and Cathy J. Robinson

Politics of Renewable Energy in China
Chen Gang

Politics of Renewable Energy in China

Chen Gang

East Asian Institute, National University of Singapore

NEW HORIZONS IN ENVIRONMENTAL POLITICS

Cheltenham, UK • Northampton, MA, USA

Published by
Edward Elgar Publishing Limited
The Lypiatts
15 Lansdown Road
Cheltenham
Glos GL50 2JA
UK

Edward Elgar Publishing, Inc.
William Pratt House
9 Dewey Court
Northampton
Massachusetts 01060
USA

A catalogue record for this book
is available from the British Library

Library of Congress Control Number: 2018959285

This book is available electronically in the **Elgar**online
Social and Political Science subject collection
DOI 10.4337/9781788118156

ISBN 978 1 78811 814 9 (cased)
ISBN 978 1 78811 815 6 (eBook)

Typeset by Servis Filmsetting Ltd, Stockport, Cheshire
Printed and bound in Great Britain by TJ International Ltd, Padstow

For China's sustainable development

Contents

Figures

Tables

1. Introduction

Environmental pollution is a global issue that most fast-growing economies have to face, relating to many factors such as economic structure, technological level, political systems, governance capacity, institutional building, as well as public awareness and social participation. China's economic miracles over the past three decades have imposed enormous pressures on the country's already worsened environment and scant resources, with the mounting ecological problems, such as emissions of greenhouse gases, urban smog, water pollution and shortages, soil contamination, desertification, and loss of biodiversity, having caught the intensive attention of the Chinese government, and the domestic public and international community. As the world's largest carbon polluter, generating about 29 percent of global carbon dioxide emissions (PBL Netherlands Environmental Assessment Agency 2016, p. 5), China has been playing a paradoxical role in global climate political economy for years, by stubbornly refusing to accept any mandatory emission-cutting obligations, yet ironically becoming the world's leading developer of renewable energies and carbon trading business. Through enhancing its capacity for environmental governance, the Chinese authorities have taken concrete steps in curbing pollution, with environmental conservation and low-carbon tasks being given the highest platform in the political agenda of the ruling Communist Party of China (CPC).

Nevertheless, a society's ability to identify and resolve environmental problems is not merely based on the knowledge and resources embedded in its bureaucracy and legal framework (Weidner 2002, pp. 1340–68). Until now, China's ecological protection and low-carbon effort has been mainly a state-led process, which has been severely restrained by the existing implementation deficit in environmental governance and the inability of the administration to monitor and reduce pollution in this vast nation. The presence of social actors who can act as advocates for the environment and the integration of these non-governmental forces in processes of planning and policy-making can substantially enhance the opportunities for ongoing environmental transition (Jänicke 1996, pp. 71–85). The rising civic activism in an era of social media and fragmentized policy-making has been constantly confronting the party-state's non-participatory approach managed through a top-down apparatus. The political interactions among central and local bureaucracies, interest groups, and societal forces have constrained the country's ambition in developing renewable energy.

The former US vice president Al Gore said in his 1993 bestselling book, *Earth in the Balance*, that 'as the dramatic environmental problems in Eastern Europe show, freedom is a necessary condition for an effective stewardship of the environment' (Gore 1993, p. 179), highlighting the nexus between democratic political institutions and the creation and implementation of environmental regulations. Yet in a new context of the global climate change crisis, some scholars have noticed that authoritarian regimes on some occasions might make better environmental managers vis-à-vis wealthy democracies' generally lackluster responses to global environmental challenges, with the concept of authoritarian environmentalism (Josephson 2004; Purdy 2010, pp. 1122–209; Gilley 2012, pp. 287–307) being developed to explain the phenomenon.

Authoritarian environmentalism has been generally interpreted in two dimensions (Beeson 2010, pp. 276–94): the first is a decrease in individual liberty that prevents individuals from engaging in unsustainable behavior and compels them to obey more sustainable policies; the second is a policy process that is dominated by a relatively autonomous central state, affording little or no role for social actors or their representatives. East Asian authoritarianism has been hailed by Shearman and Smith as a model for authoritarian environmentalism that stresses limits on individual freedoms as well as a policy process in the exclusive hands of an autonomous state. Some of the remarkable achievements made by China in curbing environmental degradation and pollution have proved that authoritarian environmentalism can be efficient and effective in fighting pollution in many ways (Shearman and Smith 2007; Chen 2009; Gilley 2012, pp. 287–307).

In recent years, the emergence of China as the world's largest producer of renewable energies, including wind and solar photovoltaic power, has been hailed by some observers as a model for authoritarian environmentalism, a policy process dominated by a strong central state that prevents businesses and localities from engaging in unsustainable behavior and compels them to obey more sustainable policies. Scholars amazed at China's renewable miracles started to challenge climate mitigation strategies prevailing in many industrialized countries that are profoundly dependent on democratic political environments and governance modes that explicitly require the strengthening of local public participation (Few et al. 2007; Devine-Wright 2012; Mathews and Tan 2015; Chen 2016). Mathews and Tan (2015, p. 1) praised China's renewable energy revolution as being 'the world's first case of a country breaking free of carbon lock-in by building its own renewable energy industries – building energy security through manufacturing.' Chen (2016, p. 8) attributed China's renewable leap forward to the government's 'hybrid strategies for environmental protection measures that combine a cluster of developmental state features in an attempt to partially incorporate a modified, state-led strategy based on ecological modernization.' China's centralized way of developing renewable energy and related industries also questions existing consensus in international documents such as the Brundtland Report (World Commission on Environment and Development 1987) and the Organisation for Economic Co-operation and Development (OECD) (2002) report that emphasized decentralized and democratic mechanisms as a necessary environmental governance condition for desired policy outcomes.

Research claims that centralization occupies an important place in the governance template used to diffuse renewable energy in China (Chen 2016, p. 9). As a non-liberal state with a centralized and top-down development model, China's leap forward in renewable energy application seems to contradict existing orthodox governance patterns in which sustainable development targets can only be achieved in a decentralized, participatory, and bottom-up environment. China's surprising progress in the low-carbon industry has been deemed as a case in point to support authoritarian environmentalism, which has been accusing democracy and the free market of promoting consumerism and individualism that exacerbate the unsustainability issue (Ophuls 1977; Methmann 2010). Most of the above literatures point to the same argument – the authoritarian Chinese regime is gifted in developing renewable alternatives because it is more capable of overcoming resistances from domestic interest groups, especially those in conventional energy sectors, than are most democracies. Besides, the party-state is well-known for drafting strategic development plans with five-year or even longer-term policy targets, which many believe helps solve the market failure in the clean energy sector where ecological benefit or cost is often not fully priced in.

Even in the authoritarian party-state, however, the Chinese regime has never been monolithic, with numerous interest groups co-existing and vying for greater influences on the extensive and fragmented bureaucracy at various levels. Modern energy politics must be understood in terms of the relative influence of numerous groups contending for protection or promotion by the government (Chubb 1983, p. 18). Evidently, China's energy policy-making does not fit in the rational decision-making model where parochial interests often give way to overall national interest. The concept of 'fragmented authoritarianism' (Lieberthal and Oksenberg 1988; Lieberthal 1992) used to describe China's decision-making in its early stage of reform is becoming more convincing for viewing today's energy policy formation and implementation in China. An authoritarian centralist state in theory, China has witnessed the emergence of de facto federalism and pluralism in practice in the past four decades of reform and opening up. In the new policy context of growing pluralization, decentralization, and fragmentation, fast-growing economic stakes have led to a subsequent increase in the number and types of pressure groups involved in energy policy-making. This is likely to expand substantially to include more ministries at the national level, big business entities, media, and non-governmental organizations (NGOs), among other entities and individuals.

The theoretical puzzle concerning interest groups and bureaucracy is not how to approach these organizational forms independently, but how to understand the relationship between them (Chubb 1983, p. 10). The outcome of China's renewable energy development is largely being shaped by the dynamic interactions among various interest groups, central–local relationships and the central government's energy policy priorities – these have been identified as the three determinants that will decide China's renewable energy future. Like many other countries, China's energy policy-making is fueled by the country's overall economic growth targets. The central government's renewable energy policy prioritizes the dual goals of meeting China's insatiable energy demand and delivering its low-carbon commitment in the context of climate change and industrial pollution. Though ambitious, the country's renewable targets are often unrealistic and are not supported by other energy interest groups or local governments. This has led to huge waste and inefficiency of renewable energy projects, with consequences of exorbitant overcapacity and curtailment rates in local wind and solar projects. The void of effective civic supervision and lobby activities impedes the enforcement of China's low-carbon energy ambition and other environmental initiatives, as political and business elites, who benefit from patronage and income from natural resource rents and favorable policies, often have little incentive to engage with citizens and to build effective public authority over energy and environmental issues.

China has an electrification rate of close to 99 percent (Urmee et al. 2009), with two-thirds of the total supply coming from carbon fuels, mostly coal. Retired Chinese Premier Li Peng, in his three-volume *Diaries of Power Industry* (Li 2005) and the *Diaries of Three Gorges Dams* (Li 2003), provides rare revelations about detailed operations and conflicts within China's fragmented power industry bureaucracy. As an expert on China's hydropower industry himself, Li witnessed the merger of the Ministry of Power Industry and the Ministry of Water Resources in 1982 (Li 2005, p. 398) to solve divergences between coal-fired power sections and hydropower departments, and the establishment of the State Council Three Gorges Project Construction Committee in 1992 (Li 2003, p. 159) to coordinate inter-agency issues concerning the construction of the world's largest hydropower project. Li's first-hand description of administrative affairs within China's electricity bureaucracy implicitly revealed that inside Chinese energy politics, under its unified and authoritarian façade, was the intensified power struggle and lobbying activities conducted by different stakeholders within the state apparatus. As a top leader in the CPC, Li chose to focus on consensus rather than divergences in the compiling of his diaries, but he failed to give convincing explanations as to why the State Council decided to abolish the leading panel on the preparation work of the Three Gorges Project in 1986 (Li 2003, p. 91), or why the then CPC general secretary Hu Yaobang was determined to push forward the merger of the Ministry of Power Industry and Ministry of Water Resources in 1982 (Li 2005, p. 399).

These unanswered questions may still link to today's Chinese energy and environmental politics, which remain authoritarian and centralized on the surface, but in reality are becoming much more fragmentized and decentralized in the context of further marketization and growing pluralism. This book focuses on ongoing interactions among various interest groups and stakeholders in China's low-carbon energy development, with special emphasis on their connections with policy-makers and their vested interests in state-dominated sectors. Given the profound impact on a broad set of interest groups in the party-state, the energy policies are often the outcomes of bureaucratic politics and inter-agency competitions within the regime rather than reasoned debates. Besides their long-existing political connections with top leadership, the fast-expanding profitability and assets of energy vested interests in the past have dramatically reinforced the economic and political clout of these groups within the bureaucracy. Energy companies, either state-owned or private, were already influential interest groups in the command economy era and this has only increased due to the vast amounts of wealth generated since then.

The Chinese central government's top-down effort to promote low-carbon power generation has encountered severe challenges in many provinces where local renewable power projects are facing high curtailment rates, and subnational governments and grid dispatchers in favor of vested coal interests are blocking renewable power generation and transmission. With an emphasis on China's renewable energy policy, and related enforcement issues and local politics concerning wind and solar power generation, this book examines to what extent China's centralized and top-down approach has been effective in coercing local actors to reach policy targets in an increasingly decentralized yet still non-democratic political context. Without marketization in the energy industry or the participation of local environmental activism, the central government does not have the capacity to overcome such local resistances. This book reviews the problems and progress in the politics of China's renewable energy development. It analyzes the factors in China's governance and political process that restrain its capacity to develop low-carbon alternatives. Focusing on the different institutional imperatives the authoritarian rulers needed to address in different stages, the book aims to make unique contributions to new institutionalist studies of authoritarianism that help to explain China's institutional capacity and weaknesses in promoting renewables and fighting climate change.

Although a couple of books on similar topics (Zhang 2011; Mathews and Tan 2015; Chen 2016) have been published in recent years, the local politics of China's renewable energy, which has been attracting growing public and scholarly interest in the new context of climate change and social media, is severely understudied. As the world's largest renewable energy producer, China has been promoting wind and solar power generation in a top-down manner that reinforced government dominance and interference in the domestic energy market and reshaped both vertical and horizontal relationships among various stakeholders in China's energy realm. Most existing studies have failed to capture the latest institutional and norm changes related to China's leap forward in developing renewable energy that reinforced central government's dominance in the fragmented local energy markets and reshaped central–local and state–society relationships in energy development. Instead of the generalized approaches adopted by most researchers that emphasize bureaucratic apparatus and policy incentives, this book focuses more on specific institutional reforms and underlining political and socioeconomic implications in this realm, with detailed description and analysis of the complicated interactions among China's central and local governments, energy companies, civic forces and other interest groups.

As China's politics and society are becoming more pluralistic and fragmented, the book is designed to help the reader better understand the dynamic relationships among various local interest groups and civic forces in promoting different kinds of energies. Although some research has touched on the emerging civic voices of the social media era, few studies have discussed the interaction among different energy stakeholders within and outside the state apparatus, which cover a wide range of institutional players, such as the central government, local governments at various levels, traditional energy producers, renewable energy producers, grid companies, environmentalist groups, official media and social media. This book draws on vivid case studies and in-depth analysis to reveal the complexity of China's formal and informal politics related to local renewable energy development, in which all these interest groups and social organizations interact with one another in a new institutional environment.

Authoritarian environmentalists tend to use China's great leap forward in wind and solar power generation as plausible evidence to challenge the pluralistic and participatory rules widely recognized in the environmental governance of most liberal democracies, which emphasize that a democratic, decentralized and bottom-up governance model provides the only possible pragmatic approach for dealing with modern environmental crises. Having seen a distinct tendency in recent years towards a centralized, top-down attitude to local governance of the renewable energy industry, some believe that such policy implementation that relies on the unusual ability of related coercive mechanisms to control verticality is the key to China achieving remarkable renewable targets.

China's better-than-expected capability in developing low-carbon energies has been seen as the outcome of its hybrid strategies that combine the developmental state and ecological modernization approaches, with local compliance being greatly enhanced through emerging formal, corporatist institutions. Authoritarian environmental research on the principal–agent relationship between central and local governments tends to believe in the omnipotent role of central government as the principal in guiding alternative energy choices within the broader governance networks, challenging the de facto federalism theory that always questions the ability of the authoritarian center as the principal to monitor or control local agents' behavior.

Given the recent predicament encountered by renewable energy projects in many Chinese provinces, the theory of authoritarian environmentalism is becoming less convincing when applied to the understanding of China's renewable energy reality. This monograph, through the analysis of such local impediments from certain interest groups, self-contradicting policy priorities, and relevant enforcement issues concerning wind and solar power generation, examines the extent to which China's centralized and top-down approach has been effective in coercing local actors to reach policy targets in an increasingly decentralized but still non-democratic political context. In reality, the central government's top-down effort to promote low-carbon power generation has been increasingly facing severe resistances in many provinces where local renewable power projects are witnessing high curtailment rates, and subnational governments and grid dispatchers in favor of vested coal interests are blocking renewable power generation and transmission. Without participation of local environmental activism, the central government does not have the capacity to overcome such local resistances.

Chapter 2 takes a close look at China's mercantile strategy to boost renewable sectors. Considering the exorbitant costs and risks involved in developing low-carbon energy, China's emergence as the world's largest player in renewable energy generation and equipment production has been closely related to the government's robust policy support and heavy subsidies under the Chinese model of state capitalism. Since the global financial crisis of 2008, when China's export-oriented manufacturing was hit hard by sluggish external demand, the government has regarded renewable energy equipment sectors as emerging industries of strategic importance that need strong policy support. Such a mercantile strategy subsequently led to serious overcapacity and trade disputes, forcing the government to stimulate domestic demand through subsidizing local power generation projects to absorb redundant manufacture capacity.

Chapter 3 focuses on China's centralized and top-down approach of developing renewable energy, which has been tagged as a model for authoritarian environmentalism, a policy process dominated by a strong central state that prevents businesses and localities from engaging in unsustainable behavior and compels them to follow more sustainable policies. In each stage of its renewable sector development, the change in Chinese central leadership's policy priority has always been triggered by both international and domestic pressures. The statutory and enforcement flaws reflected in the Renewable Energy Law resulted from improvised policy priority change from top leadership, unpreparedness of central and local enforcement apparatus and resistances from within the bureaucracy and various interest groups. Many of the central government's supportive policies in the beginning proved to be unrealistic and unattainable, only starting to take effect when market conditions started to become more mature and interest groups became more cooperative. This chapter examines the innate weaknesses of such a centralized and top-down approach in fostering renewable energy development.

Chapter 4 analyzes local geographic and industrial barriers that have been challenging the implementation of national renewable energy policies. China's wind and solar resources are concentrated in the vast but sparsely populated northwestern and northern areas, exacerbating the imbalanced distribution of energy resources that constantly afflicts the economically vibrant eastern (coastal) area with energy shortage. The overconcentration of conventional coal, hydroelectricity, and emerging wind and solar resources in the western part of Chinese territory has imposed enormous pressure on the country's transmission networks. Despite central government's effort to promote renewable energy, such intentions have resulted in huge waste and overcapacity in the northwest and northern provinces. This chapter discusses why high curtailment rates and overcapacity are inevitable in remote areas.

Chapter 5 discusses policy priorities advocated by different interest groups, and how such interaction would affect the implementation of China's renewable policy. China's weakness in enforcing national energy policies is subject to the level of coordination within three kinds of relationships, namely the relationship between central government agencies, the relationship between central and local governments, and the relationship between government agencies and state-owned enterprises (SOEs). This chapter discusses policy priorities advocated by different interest groups, and how such interaction affects the implementation of China's renewable policy.

Chapter 6 continues the discussion on different interest groups in the low-carbon energy sector. Mitigation imperatives coupled with energy shortage have pushed China to encourage aggressive construction of new hydropower and nuclear plants, in addition to its impressive input into such established renewable energies as wind and solar power. Facing fierce competition among different low-carbon alternatives, the Chinese government has been adjusting subsidy amount, on-grid tariffs, and other financial incentives from time to time in supporting various non-fossil fuels. Compared with other democracies, the authoritarian one-party regime in China faces much less resistance from civil society when pushing forward those hydropower and nuclear mammoths that may pose potentially serious threats to the local ecology.

Chapter 7 discusses energy policy priorities in China's Five-Year Plans (FYPs). The core of all energy-related policies of Chinese top planners includes maximizing/diversifying energy supply and controlling/conserving energy demand. Scrutiny of changes in policy priorities in China's energy FYPs contributes to the understanding of how the central authorities have adjusted energy policy targets in response to new industrial and market conditions as well as concerns from different interest groups. The scrutiny of FYPs reveals that the prioritization of sufficiency over efficiency, a vestige from China's pre-reform communist ideology of planning economy, and a consequence of the slow process of energy marketization and privatization, is still deeply rooted in China's energy apparatus, and will continue to dominate policy-making for a long time.

Chapter 8 examines how China's energy policy priorities have restrained the authoritarian state from tapping the full potential of its low-carbon resources. China's renewable energy strategy is a natural extension of the country's preferred 'no-regret' strategy that emphasizes mitigation actions providing fringe benefits such as economic growth and employment to the country, regardless of whether the threat of climate change is real. On the surface, China's coal-dominant energy structure puts it in an awkward dichotomy between energy supply and environmental protection, but in reality, the authoritarian state's obsession with high economic growth, powered by relentless domestic energy production, is the rooted cause for the economic model featuring inefficient energy utilization and high-polluting emissions. Such a development-centric ideology of government and lack of environmental activism at the civic level would prevent China from making full use of the renewable energy facilities installed across the country.

The final chapter summarizes the discussion on interaction among China's renewable energy policy priorities, relevant interest groups and the central–local relationship. Focusing on the political and institutional factors leading to China's redundant renewable power capacity, the book argues that China's state capitalism has an innate tendency to focus on the 'supply side' instead of 'demand side', which differs from a market-driven economy and results in enormous industrial capacity subsidized by the state yet detached from real market demand. To fundamentally solve the huge waste problem associated with excessive investment and overcapacity, China has to transform its power sector from a top-down and state-driven system to a new version driven by domestic demand and with less intervention from governmental apparatus at various levels.

REFERENCES

Beeson, M. (2010), 'The coming of environmental authoritarianism', *Environmental Politics*, 19 (2), 276–94.

Chen, G. (2009), *Politics of China's Environmental Protection: Problems and Progress*, Singapore: World Scientific.

Chen, G. C.-F. (2016), *Governing Sustainable Energies in China*, New York: Palgrave Macmillan.

Chubb, J. E. (1983), *Interest Groups and the Bureaucracy: The Politics of Energy*, Stanford: Stanford University Press.

Devine-Wright, P. (ed.) (2012), *Renewable Energy and the Public: From NIMBY to Participation*, London: Routledge.

Few, R., K. Brown, and E. L. Tompkins (2007), 'Public participation and climate change adaptation: avoiding the illusion of inclusion', *Climate Policy*, 7 (1), 46–59.

Gilley, B. (2012), 'Authoritarian environmentalism and China's response to climate change', *Environmental Politics*, 21 (2), 287–307.

Gore, A. (1993), *Earth in the Balance: Ecology and the Human Spirit*, New York: Plume.

Jänicke, M. (1996), 'Democracy as a condition for environmental policy success: the importance of non-institutional factors', in W. M. Lafferty and J. Meadowcroft (eds), *Democracy and the Environment: Problems and Prospects*, Cheltenham: Brookfield, pp. 71–85.

Josephson, P. R. (2004), *Resources under Regimes: Technology, Environment, and the State*, Cambridge, MA: Harvard University Press.
Li, P. (2003), *Diaries of Three Gorges Dams* (*Li Peng Sanxia Riji*), Beijing: Zhongguo Sanxia Chubanshe.
Li, P. (2005), *Diaries of Power Industry* (*Li Peng Dianli Riji*), Vols 1–3, Beijing: Zhongguo Dianli Chubanshe.
Lieberthal, K. (1992), 'Introduction: the "fragmented authoritarianism" model and its limitations', in K. Lieberthal and D. M. Lampton (eds), *Bureaucracy, Politics, and Decision Making in Post-Mao China*, Berkeley, CA: University of California Press, pp. 1–30.
Lieberthal, K. and M. Oksenberg (1988), *Policy Making in China: Leaders, Structures and Processes*, Princeton, NJ: Princeton University Press.
Mathews, J. A. and H. Tan (2015), *China's Renewable Energy Revolution*, New York: Palgrave.
Methmann, C. P. (2010), 'Climate protection as empty signifier: a discourse theoretical perspective on climate mainstreaming in world politics', *Millennium – Journal of International Studies*, 39 (2), 345–72.
Organisation for Economic Co-operation and Development (OECD) (2002), *Governance for Sustainable Development: Five OECD Case Studies*, Paris: OECD.
Ophuls, W. (1977), *Ecology and the Politics of Scarcity*, San Francisco, CA: W. H. Freeman.
PBL Netherlands Environmental Assessment Agency (2016), *Trends in Global CO_2 Emissions: 2016 Report*, The Hague: PBL Netherlands Environmental Assessment Agency.
Purdy, J. (2010), 'The politics of nature: climate change, environmental law, and democracy', *Yale Law Journal*, 119, 1122–209.
Shearman, D. J. C. and J. W. Smith (2007), *The Climate Change Challenge and the Failure of Democracy*, Westport, CT: Praeger.
Urmee, T., D. Harries, and A. Schlapfer (2009), 'Issues related to rural electrification using renewable energy in developing countries of Asia and Pacific', *Renewable Energy*, 34 (2), 354–7.
Weidner, H. (2002), 'Capacity building for ecological modernization: lessons from cross-national research', *American Behavioral Scientist*, 45 (9), 1340–68.
World Commission on Environment and Development (WCED) (1987), *Our Common Future*, Oxford: Oxford University Press.
Zhang, Z. (2011), *Energy and Environmental Policy in China: Towards a Low-Carbon Economy*, Cheltenham, UK and Northampton, MA: Edward Elgar.

2. China's mercantile strategy to boost renewable sectors

Amidst increasing global concerns about climate change and energy security, the booming renewable energy industry has emerged as one of the major business opportunities of the twenty-first century. In research by the PEW Charitable Trusts, these clean energy projects include all biomass, geothermal, and wind generation projects of more than 1 MW, all hydro projects of between 0.5 and 50 MW, all solar projects of more than 0.3 MW, all marine energy projects, and all biofuel projects with a capacity of one million or more liters per year (PEW Charitable Trusts 2010, p. 41).

Under the mercantile model of state capitalism, in the initial stage of incentivizing renewable sectors China paid more attention to export-oriented production of wind turbines and solar panels rather than the costly renewable power generation and consumption on its homeland. Such an asymmetrical development strategy subsequently led to serious manufacturing overcapacity and chronical trade disputes with the United States and European Union (EU), which forced the government to gradually stimulate domestic demand through subsidizing local power generation projects to absorb redundant manufacture capacity. China's transition from subsidization of export-oriented manufacturing to domestic demand stimulation was also state-driven, proven to be unnatural and painful with unintentional consequences of overcapacity in domestic renewable power generation projects.

OVEREXPANSION OF CHINA'S RENEWABLE EQUIPMENT MANUFACTURING AFTER GLOBAL FINANCIAL CRISIS

China's industrial planning has included a focus on increasing Chinese capacity and production of wind turbines, solar cells and modules, using state incentives, subsidies, and tariffs to dominate the global supply chain. China institutionalized its efforts to develop clean energy capacity by passing the groundbreaking Renewable Energy Law in 2005, which was amended and reinforced in 2009 with stronger punitive articles included.

Under the framework of Renewable Energy Law and national green targets set by the Medium- and Long-Term Development Plan for Renewable Energy (2007), detailed institutional incentives and disincentives have been designed by various layers of the government to promote renewable energy development in China (Chapter 3). Since the global financial crisis of 2008, when China's export-oriented manufacturing was hit hard by sluggish external demand, the government has regarded renewable energy equipment sectors as emerging industries of strategic importance (*zhanlüe xinxing chanye*) that need strong policy support. The government's decisive actions to support and subsidize clean-energy industry have received remarkable paybacks. China's manufacturing sector for producing wind turbines, solar panels, and modules has been booming as a new engine of economic growth in the aftermath of the financial crisis. Thanks to strong state support, China's production capacity of these renewable power generation equipment and parts quickly surpassed bellwethers such as the United States, Japan, and some European countries, dominating the global market by the largest shares and bringing down the international prices by large margins.

Prior to 1994, there were only a few demonstration wind farms fully financed by the government and, from 1994 to 2003, in-grid wind power was first developed as a new energy source for electrical power generation in China (Yang and Pan 2010a, p. 5). To import technology from foreign companies and to establish a high-quality Chinese wind turbine generator sector, the former State Development and Planning Commission (SDPC), predecessor of the National Development and Reform Commission (NDRC), initiated the 'Ride the Wind Program' (*chengfeng jihua*) in 1996, leading to the establishment of two joint ventures, NORDEX (Germany) and MADE (Spain), which effectively introduced 600-kW wind turbine generator manufacturing technology to China (Ecoworld 2008). China's wind turbine industry, a new growth point that is immune from the impacts of the global economic downturn, has witnessed galloping expansion since 2008, bringing down turbine prices by as much as 20–25 percent in Western markets and more than 35 percent in China between 2008 and 2012 (Renewable Energy Policy Network for the 21st Century 2013, p. 54). Among the world's top ten turbine manufacturers that captured 77 percent of the global market in 2011, four hailed from China.

Following state-directed initiatives, China's share of global solar cell production also surpassed frontrunners such as the United States and Japan, skyrocketing from 7 percent in 2005 to 61 percent in 2012. China now dominates global supply chain capacity, accounting for nearly 70 percent of total planned global capacity expansions announced in the first half of 2017. Even in 2017, China still produced 60 percent of the world's solar cells and 71 percent of solar modules (US Trade Representative 2018). Because the production of solar cells and modules is a capital and tech-intensive industry in which China had no comparative advantage a decade ago, its swift success in snatching up the high ground from the control of formidable peers in the United States and Europe shocked many industrial observers with the expectations that the Western world would alleviate trade imbalance through the development of such high-tech sectors.

The comparative advantages that China has accumulated during the short period in which it has been producing wind turbines and solar modules are outstanding, and defy standard explanations of China's manufacturing competitiveness based on cheap labor, good infrastructure, or the alleged undervaluation of the Chinese currency. In the nascent solar photovoltaic (PV) manufacture industry that has emerged amidst global concerns about climate change and energy security, labor formed just a tiny portion of manufacturing costs (Haley and Haley 2013, p. 1), while capital, technology, land, environmental regulation, and scale of production are more crucial factors in determining competitiveness. EU ProSun (2013, n.p.), the main solar panel manufacturers' lobby group, attributed the lower Chinese prices, which contradicted the reality that European production costs are lower than in China, to 'illegal state subsidies' and 'widespread dumping as proven by the EU and US investigations.'

In its infant stage of solar PV manufacture, China was weak in terms of technologies for producing the raw material for solar PV, polycrystalline silicon. Despite the establishment of a number of research institutes and universities dedicated to research and development (R&D) for solar PV between 1986 and 2005, advancement of PV technologies was slow in China (Yang and Pan 2010b, p. 14). In the early days, China did not have competitive PV-related knowledge resources that resided in universities, government research institutes, private research facilities, business and scientific literature and market research databases (Chen 2015, p. 101). China had to import 95 percent of its PV raw materials due to a lack of advanced technologies for producing polycrystalline silicon, and it exported 95 percent of its finished products due to an underdeveloped domestic market (Li 2010, p. 38; Chen 2015, p. 98).

In the solar PV industrial chain that includes production of crystalline silicon, silicon ingot and wafer, solar cells, modules, and arrays, China was only able to manufacture a small proportion of global silicon material in the upstream while it accounted for about half of global production capacity for cells and modules in the downstream. In other words, in the nascent stage, China was weak in the upstream of the PV industrial chain, namely silicon production, while it was strong in the labor-intensive downstream, including cell production and PV module encapsulation (Table 2.1).

Table 2.1 China's manufacturing capability in the solar PV industrial chain (2008–2009)

Industrial Chain	Silicon	Ingot/Wafer	Cells	Modules	PV Arrays
Features	Capital-intensive; high technical threshold; long payback period; high energy consumption; and poisonous emissions.	High technical barrier; decrease in the thickness of silicon wafers is an effective way to reducing the cost of PV cells.	Large-scale production is crucial; medium-level capital and technical threshold.	Low capital and technical threshold; labour-intensive; short payback period; and fierce competition.	Low technical threshold but comprehensive use of combined technologies is required.
Manufacturing Competitiveness in China	Weak	Medium	Strong	Strong	Strong

Source: Yang and Pan (2010b, p. 25).

Production of high-purity polycrystalline silicon, the basic feedstock of solar cells, has always been the cause of China's bottleneck situation in the industrial chain (Li and Ma 2009). Despite its rich silicon reserves, China was lagging behind in the processing of silicon purification at the international level in the mid-2000s; much of its high energy consumption and emission of poisonous waste was owing to backward technologies and lack of strict regulation. Although challenges such as low efficiency and pollution also existed in other countries, China's situation was more extreme. As regards ingot and wafer production, the second stage of the industrial chain, China's silicon ingot slicing industry was relatively mature at the international level, with thinner wafers developed and thus less silicon feedstock needed. However, since the status of silicon ingot and wafer production was still largely dependent on the silicon feedstock, there was still a significant gap between China and some industrialized countries in terms of wafer production skills. In contrast to its weak competitiveness in the upstream, China was strong in the downstream solar cell manufacturing and module encapsulation, in which the advantage of low labor costs is particularly important.

The high energy consumption and emission of poisonous waste in the production of high-purity polycrystalline silicon have added both economic and ecological costs to China's PV manufacturing efforts. China's early solar production was beset by problems similar to those that India's PV production was facing: high levels of pollution and high energy costs, fragmented manufacturing capacity, lack of scale, an underdeveloped local capital market, and lagged R&D. The explanation for China's later triumph and India's stagnation in PV production can be found through the study of factor distortion under the framework of state capitalism. Historically, India has had more of a free-market orientation than China (Shilling 2012), and its state sector is more fragmented, decentralized, and less interventionist compared with China's government apparatus. Sharing a similarly humble industrial basis, China's PV manufacturing has outrun India's largely due to the interventionist role played by a strong Chinese government.

China is a game-changer in global solar PV manufacturing, as its revenue share in the world market surged from an insignificant portion before 2005 to more than 50 percent in 2011 and 2012. China's existing production capacity for solar panels was about 150 percent of global real demand in 2012, and in just four years, between 2009 and 2012, the international price of solar panels had been cut by more than 75 percent, largely due to an aggressive capacity build-up in China (Renewable Energy Policy Network for the 21st Century 2013, p. 47).

Both the EU and the US authorities have accused China of illegal state subsidies and widespread dumping in the drive for industrial competitiveness, although this does not take into account the reality that production costs in the EU and US were actually lower than in China. As noted, compared with other industrialized nations, China is well-known for its advantages in the supply of cheap labor; but in solar PV manufacturing, labor accounts for a small part, less than 10 percent of total costs (Chen 2015, p. 101). Both the EU and US authorities found that the Chinese government's generous subsidies had helped its PV producers to gain comparative advantages by reshaping the country's factor conditions, which included human resource input, infrastructure, knowledge resources, and the capital market. In 2011, US Commerce found that China had subsidized its solar cell and module producers, and that those producers were selling their goods in the United States for less than their fair market value to the detriment of US manufacturers (US Trade Representative 2018).

FROM SUBSIDIZATION OF EXPORT-ORIENTED MANUFACTURING TO DOMESTIC RENEWABLE POWER GENERATION

Unlike its persistent interest in hydro and nuclear for the sake of energy security, China turned its attention to wind and solar power generation much later, largely in response to the new context of climate change and redundant manufacture capacity of renewable equipment. As a latecomer to renewable power generation, China managed to emerge swiftly as the world's major renewable energy powerhouse with strong national policies aimed at incentivizing the use of solar and wind (Chen 2012, p. 46; Salim and Rafiq 2012, p. 1051). After four years of implementing the Renewable Energy Law, the central government found that, although the country had rapidly increased its installed capacity of renewable energy, much of this capacity was not promptly connected to the grid and that not all power being generated was being purchased as required by the law.

When the Renewable Energy Law was amended in 2009, the top legislature required electricity grid companies to buy all the power produced by renewable energy generators and authorized the State Council energy department, in conjunction with the state power regulatory agency and the State Council finance departments, to determine the percentage of the quantity of electricity generated from renewable energies in the total quantity of electricity generated during thc planned period (PRC Amended Law on Renewable Energy 2009, Article 14), a regulation system similar to the Renewable Portfolio Standards (RPS) adopted by some US states. The amendment made it clear that power enterprises refusing to buy power produced by renewable energy generators would be fined up to an amount double that of the economic loss of the renewable energy company (PRC Amended Law on Renewable Energy 2009, Article 29). To encourage grid companies to accept more power generated from renewable sources, the amendment waived the value-added and income taxes levied on grid companies on revenue (the tax rate is about one-third of the revenue) generated from the surcharge on the retail power tariff for supporting renewable energy (Renewable Energy Policy Network for the 21st Century 2009, p.20).

The landmark policy came in 2003 when the NDRC launched the 'Wind Power Concession Program' (*fengdian texuquan xiangmu*) to build large-capacity wind farms and achieve economies of scale through reducing the in-grid wind power tariff. Under the program, the power grid company signs a long-term power-purchase agreement with the wind power project investor and agrees to purchase the prescribed amount of electricity generated by the project, whose capacity must reach 100 MW. Investors and developers of wind farms are selected through a competitive bidding process that determines the in-grid tariff. All end-users of the grid's electricity share the tariff increase due to wind power purchase. As incentives, the government waives the import customs tariff and value-added tax on the equipment and accessories. The 'Wind Power Concession Program', which minimizes the risks to investors through government-guaranteed power-purchase agreements, pushed China's wind power industry into a fast-growing stage between 2005 and 2009,with the country's installed wind power capacity more than doubling annually in that four-year period. By the end of 2005, China had built 59 wind farms with 1,854 wind turbine generators and a 1,266 MW in-grid wind power installed capacity, ranking it at number ten globally. In 2009, China's installed wind power capacity reached 25.81 GW, about 20 times that in 2005, ranking number two after the United States. From 1997 to 2004, however, China's installed wind power capacity maintained only a relatively low level of growth rate, fluctuating between 16.3 percent and 28.4 percent (Yang and Pan 2010a, p.6).

Owing to massive producer subsidies, research grants, tax rebates, low-interest loans, and cheap land under the model of state capitalism, China's manufacturers of wind turbines and solar panels, accounting for the largest share of the world market, have found themselves caught in a plight of overcapacity, slump in selling price, and exacerbated trade rows. Massive investment and capacity built up in China led to serious oversupply in the international market and drove down turbine and module prices substantially (Renewable Energy Policy Network for the 21st Century 2013, p.47). As a result of China's fast-growing market share, both the EU and the United States launched anti-dumping and anti-subsidy investigations of the country's wind turbine and solar panel shipments, which became the biggest commercial disputes of their kind in history.

The costs of operating and maintaining wind farms dropped significantly due to increased competition among contractors and improved turbine performance. The domestic manufacturing boom has justified the Chinese government's approach to promote inland wind-generated power, which became gradually cost-competitive vis-à-vis conventional power and thus needed fewer subsidies on a per kilowatt-hour basis. In the context of climate change and energy scarcity, the country has been formulating one of the world's most aggressive strategies to promote its low-carbon energy production from renewable sources. From 2005 to 2010, compared with solar (mainly solar PV) power generation, wind power was positioned at the center of renewable energy development, which had resulted in differentiated growth patterns in favor of wind. During this period, the installed capacity of wind power was 89.8 percent per annum, while solar PV's installed capacity, until 2010, remained 0.86 GW, a paltry proportion of 0.09 percent of the country's total installed capacity (China's 12th Five-Year Plan on Energy Development 2013). Since the beginning of the 12th Five-Year-Plan (FYP) period (2011–15), the Chinese government, in response to new industrial conditions of leapfrogging PV manufacturing capacity, has been adjusting its low-carbon-electricity strategy with the focus shifting from wind to solar. A reverse trend has occurred during the 12th FYP, when solar grew 122 percent but wind grew only 33 percent annually (China's 13th Five-Year Plan on Energy Development 2016). By the end of 2015, China's installed capacity of wind power and PV power generation reached 129 GW and 43 GW, respectively, both ranking the first in the world and accounting for 7.8 percent and 2.6 percent respectively of China's total installed capacity (China's 13th Five-Year Plan on Energy Development 2016).

UNINTENDED CONSEQUENCES AND LATENT RISKS

China's burgeoning environment industry faces serious problems of overcapacity and latent financial risks in this round of investment frenzy in the green market. Concerns about a national wind and solar bubble are mounting, while billions of dollars are being pumped into renewable sectors. China's commitment to renewable energy is expensive. Although costs are falling steeply through mass production, wind energy is still 20–40 percent more expensive than coal-fired power. Solar power is still at least twice as expensive as coal. Experts worry that, without state assistance or more favorable policies, many large-scale projects will not be profitable and may end up with huge losses or even in bankruptcy.

In terms of wind power development, China often faces the bottleneck problem of connecting its impressive wind power capacity to the national transmission power grid. Even with the 'mandatory grid connection' and 'renewable portfolio target' schemes institutionalized by the powerful Renewable Energy Law, oligarchic grid companies are still in an advantageous position when bargaining with numerous power plants across the country, where not all power generated can be transmitted. The conflict between the grid companies and electricity generation sectors (*wangdian zhizheng*) was entrenched long before the booming of clean energy, which has led to a huge efficiency loss on the supply side and low prices of electricity sold by power plants to grid companies. Meanwhile, most wind turbines are installed in remote areas too far away from the transmission grid and energy-intensive urban areas.

It is estimated that about 30 percent of installed wind capacity in China is lying idle and not connected to the power grid (Forbes 2009). Thanks to the government's generous subsidies and local governments' relentless pursuit of gross domestic product (GDP), overinvestment and overcapacity exist in both the wind power generation sector and wind turbine manufacturing. In just a few years, stimulated by forceful government support and attractive state subsidies, China's wind power capacity and related component manufacturing and turbine assembling sectors have undergone massive expansion with signs of overheating. As the province with the richest wind resources in China, Inner Mongolia's installed wind power capacity approached 11.4 GW by March 2011, accounting for at least 30 percent of the nation's total; however, a large portion of the electricity generated there could not be transmitted elsewhere, which has caused huge losses for local windmills (sina.com news 2011).

For many large-scale wind farms located in the remote western areas that are rich in wind power, the only chance of profit comes from state subsidies under the country's new energy strategy. If the government reduces its support considerably, both turbine installation and the sale of electricity to the grid may turn out to be loss-making and unsustainable businesses. As a result of a 'Buy Chinese' policy to favor local manufacturers, many wind turbine manufacturers have entered the sunrise industry with insufficient technology or capital. Among the 80 major Chinese wind turbine producers, only ten are capable of annual batch production of over 100 units. The remaining 70 manufacturers accounted for only 2.3 percent of the total market share in terms of new installed capacity in 2008 (Yang and Pan 2010a, p. 16). Challenges facing China's solar PV industry come from both the supply side and demand side, with over-subsidy and overinvestment having caused serious overcapacity in the industry. In the second half of 2009, China had 80,000 tons of multi-crystalline silicon production facilities under construction, posing a severe potential overcapacity when China's actual annual demand for such key material in the PV power industry was 25,000–30,000 tons. Known as the asymmetrical problem of Both-Ends-Out (*liangtou zaiwai*).

In the solar PV industrial chain that includes production of crystalline silicon, silicon ingot and wafer, solar cells, modules and arrays, China can only manufacture 18 percent of global silicon material in the upstream while it accounts for 47 and 54 percent of global production capacity, respectively, for cells and modules in the downstream (Yang and Pan 2010b, p. 14). Meanwhile, the high energy consumption and emission of poisonous waste in the production of high purity polycrystalline silicon have cast doubts on an industry labeled as green. Owning rich silicon reserves, China has witnessed a boom in silicon manufacturers and hence a production surge driven by substantial profits, but pollution and energy guzzling in the processing of silicon purification have forced the government to enact directives to discourage the low-level replication of such production lines. In the long run, the overdependence on overseas markets increases the vulnerability of China's PV industry as declining foreign demand may become a major constraint to its development.

China's overexpansion of its manufacturing capacity in renewable energy equipment like wind turbines and solar panels quickly had a substantial spillover effect in the international market, leading to serious trade disputes with Western supplies. The United States in 2010 accused China of illegally subsidizing the production of wind power equipment and asked for talks at the World Trade Organization (WTO), the first step in filing a trade case (Reuters 2010). US trade officials said Chinese manufacturers of wind turbines and related parts and components could have received several hundred million dollars in questionable government grants in 2008 under China's Special Fund for Wind Power Manufacturing. The grants appeared to violate WTO rules by requiring Chinese manufacturers to use only Chinese-made parts and components. Half a year later, China agreed to stop subsidizing Chinese wind turbine makers in the form of grants that had encouraged them to use key parts and components made in China rather than buying imports (Reuters 2011).

China's surging solar panel capacity subsequently triggered much larger trade disputes with the United States and the EU. In September 2012, The European Commission launched an anti-dumping investigation into imports of solar panels and their key components (i.e., solar cells and solar wafers) originating in China. In terms of import value affected, this was the most significant anti-dumping complaint the European Commission has received so far, as in 2011, China exported solar panels and their key components worth around €21 billion to the EU (European Commission 2012). Almost at the same time, the US Commerce Department also started to impose punitive duties on solar panel imports from China, after finding that Chinese solar manufacturers receive government support.

Anti-subsidy investigations by both the US Department of Commerce (2012) and European Economic and Trade Office (EU ProSun 2013) found that the Chinese solar PV producers had obtained government subsidies – in the form of discounts on raw materials, electricity and funding as well as export subsidy (export buyer's credits) – of about 15 percent of their sales in the US case and 11.5 percent in the EU case (Table 2.2). The European case was four to five times larger by value than the investigation by the United States, in part because the EU was the biggest export market for Chinese solar panels. In January 2018, targeting cheap Chinese products and companies moving manufacturing operations

Table 2.2 China's subsidy rates for solar manufacturers in the US and EU case (2012/2013)

	Producer/Exporter	Subsidy Rate
EU Investigation	All Chinese Solar Manufacturers	11.5%
	Wuxi Suntech Power Co. Ltd.	
	Luoyang Suntech Power Co. Ltd.	
	Suntech Power Co., Ltd.	
	Yangzhou Rietech Renewal Energy Co., Ltd.	
	Zhejiang Huantai Silicon Science & Technology Co., Ltd.	
	Kuttler Automation Systems (Suzhou) Co. Ltd.	
	Shenzhen Suntech Power Co. Ltd.	
	Wuxi Sunshine Power Co. Ltd.	
	Wuxi University Science Park International Incubator Co., Ltd.	
US Investigation	Yangzhou Suntech Power Co. Ltd.	14.78%
	Zhejiang Rietech New Energy Science & Technology Co. Ltd.	
	Changzhou Trina Solar Energy Co. Ltd. and Trina Solar (Changzhou) Science & Technology Co. Ltd.	15.97%
	All others	15.24%

Sources: US Department of Commerce (2012) and EU ProSun (2013).

into neighboring countries, the US President Donald Trump decided to impose safeguard tariffs on solar module and cell manufacturers for the next five years, starting at 30 percent and ending at 15 percent in Year 4 (PV Magazine 2018).

Intensified trade disputes with major trading partners have put the Chinese central government in awkward diplomatic positions, as its officials deem acceptance of WTO norms and support of free-trade accords as being important for enhancing its soft power and gaining respect from others. Active participation in international trade negotiations and institutions has been an important part of China's diplomatic activities since the early 1990s, effectively promoting China's international image and elevating its soft power. In addition to foreign pressure, the central government has also been facing challenges arising from mounting local debts associated with over-investment and overcapacity. Since 2012, the central government has been imposing pressures upon local governments and state banks that financed a burst of wind and solar manufacturing capacity across the country.

As with other manufacturing sectors, the local Chinese governments, from provincial to county level, provided factories along the industrial chain of PV manufacturing with various forms of subsidies, which included free or low-cost loans, tax rebates, research grants, cheap land, energy subsidies, and technological, infrastructure, and personnel support. One of the easy ways for local governments to support indigenous PV producers was to instruct local commercial banks to offer bountiful loans at low interest. By lowering the costs of capital, subsidized credit supports PV manufacturers regardless of their productivity levels, even if some of these loans may face a high risk of default. In many cases, local officials encouraged entrepreneurs to take more loans from local banks for capacity expansion, regardless of oversupply and competitiveness problems.

It was reported that in 2005, after a mere half-an-hour conversation with Peng Xiaofeng, CEO of LDK Solar (which later became the world's second-largest maker of solar wafers), Mayor Wang Dehe of Xinyu City, Jiangxi Province, immediately asked local banks to provide loans worth 200 million yuan (US$26 million) for the start-up (*Meiri Jingji Xinwen* 2012). Between 2005 and 2012, the local bank loans accredited to Wuxi Suntech Power Co Ltd, once China's largest PV manufacturer, jumped from US$56 million to US$3.7 billion, largely due to a municipal government mandate on local state-owned banks for providing low-interest loans to Suntech. From 2006 to 2011, Wuxi Suntech also received tax rebates and other forms of refund amounting to 8.65 billion yuan (about US$1.42 billion) from the government with the aim of promoting exports (*China Business Times* 2013).

This huge credit influx into domestic PV manufacturing exacerbated the existing overcapacity and debt issues. By August 2012, China's top ten PV makers had already accumulated a combined debt of US$17.5 billion, leading the whole industry to the brink of bankruptcy (*Caijing* 2012). Despite this worsening debt, local governments, to protect local GDP growth and employment, continued to grant financial aid in all forms to save plants from being closed down. Under the framework of state capitalism, state-run banks offer interest subsidies to loans for PV manufacturing and related technology R&D, a key sector that China wanted to support as a new growth point in the aftermath of the global financial crisis. Rocketing bank loans as a result of government connivance exacerbated the overcapacity and debt problems of Wuxi Suntech, which was experiencing weak sales and was on the brink of bankruptcy in 2012. Ignoring the principle of prudence, the Wuxi municipal government still gave directives to the local subsidiary of Bank of China, asking it to disregard the risk of default and grant emergency loans of 200 million yuan to the company (*China Business Times* 2013).

Besides direct fund injection, provision of cheap land is another approach often used by local governments to promote local PV production and investment. Most PV producers need sizable industrial land for assembling production lines, R&D, logistical purposes, and waste and pollution treatment, but in the start-up phase, they usually lack sufficient funds or collateral to acquire this. Local governments, through confiscating land from villagers, urban residents, or other factories at very low prices, often have abundant landbanks to support all kinds of industrial enterprises. In the start-up stage in 2005, the Jiangxi-based LDK Solar asked for 7,012 *mu* (467 hectares) of land from the Jiangxi provincial government, which at that time had no extra land-use quota for large manufacturers. Jiangxi nonetheless granted land of 15,000 *mu* to LDK Solar, most of which was newly confiscated farmland leased to LDK at very low prices (*Meiri Jingji Xinwen* 2012). To attract prominent domestic PV manufacturers to build their production lines in Heyuan City, Guangdong Province, the Guangdong provincial government not only shortened the approval process for any new PV manufacturing projects to only five working days, but also persuaded other enterprises in the same industrial park to accept land reallocation without compensation. One factory which had to concede part of its land right to incoming PV manufacturers had asked for economic compensation amounting to 40 million yuan, but following interference from the local government it eventually withdrew its claim (*Meiri Jingji Xinwen* 2012).

As with many industrial sectors in China, energy consumption is also subsidized extensively to lower costs and thus promote the competitiveness of local PV producers. As most solar PV manufacturers are electricity guzzlers, such energy subsidies in the forms of preferential electricity prices and power infrastructure support are vital for the survival of these producers in fierce competition with their domestic and international peers. Electricity prices have been tightly controlled by the Price Bureau of the NDRC, the powerful department in charge of economic plans inside the State Council, China's cabinet.

When setting power prices, the NDRC usually accommodates provincial stakeholders' interests and, in most cases, provincial officials lobby for end-users to accept only low power prices to keep their industries viable and competitive. China has set electricity tariffs lower than supply costs and lower than the tariffs in developed countries, and although electricity prices for Chinese industry appear higher than those for residential use, in practice many companies do not bear the full costs indicated by the official average figures, with subsides or 'price adjustments' permeating the system (Haley and Haley 2013, pp. 38–9). Industrial sectors consume about three quarters of the total electricity generated in China, with local economic and social concerns often propelling local governments to grant power price concessions to high-energy-consuming industries. China's State Electricity Regulatory Commission once openly criticized the Sichuan and Heilongjiang provincial governments in its annual report for subsidizing local high-energy-consuming enterprises like solar PV manufacturers and chemical factories by arbitrarily lowering the on-grid electricity prices sold to them. According to the report, between January and May 2010, the Heilongjiang provincial government lowered the selling price of energy for PV producers and other energy guzzlers from 533 yuan/1,000 kwh to 480 yuan/1,000 kwh, which helped these companies collectively save 100 million yuan after they had bought electricity of 753 million kwh at this preferential price during that period (China's State Electricity Regulatory Commission 2011, pp. 13–14).

Provincial governments sometimes broker special deals for large PV producers that have a significant impact on the local economy, or they build extra power facilities to ensure a direct electricity supply. In response to the enormous energy appetite of LDK Solar, the state-owned power company in Xinyu City also built six transformer substations with a total capacity of 1,520,000 KVA (1,000 VA), equal to the capacity of the power grid in western Jiangxi Province, for the sole use of LDK Solar (*Meiri Jingji Xinwen* 2012).

The central government, ironically, also gave technological, infrastructure, and personnel support to local solar panel producers to enhance their competitiveness in this tech- and capital-intensive industry. Yingli Solar, the world's biggest manufacturer of solar PV, is based in northern China's

Hebei Province. It has succeeded in setting up two laboratories for PV technology, with substantial technologic, financial, and personnel support from the Chinese Ministry of Science and Technology and the National Energy Commission under NDRC. The government provided these renewable energy labs with generous hardware (facilities, infrastructure, and funding) and software (policy, research staff) support, which can help local enterprises to cut the costs of R&D. Yingli Solar set up its own PV laboratory in January 2010, but a large part of its initial investment of 540 million yuan actually came from the government instead of Yingli.

China's mercantile policy targets, including increase of trade surplus, fiscal revenues, and local employment, were pursued by both the central and local governments, as well as manufacturers, commercial banks, research institutes, and other stakeholders linked to them. Such a policy priority led to oversupply of renewable energy equipment despite the country's weak domestic demand in the early stage of developing its renewable power generation. The central government, though, gradually felt political and economic pressures from both foreign and domestic dimensions as the surging manufacturing capacity not only resulted in growing trade spats, but also led to defaults on corporate bonds and local debts. Ironically, even when the central government started to rein in the built-up of production lines, the momentum could not be altered as planned because local governments and other interest groups benefiting from the local manufacturing frenzy found it hard to wean off the subsidization of the industry. Even certain central departments, such as the ministry of science and technology or the ministry of industry and information technology, were continuing to support large manufacturers. Without effective participation of environmental NGOs or other social forces, China's manufacture of renewable energy equipment has always been driven by mercantile motives rather than environmental or climate consciousness.

REFERENCES

Caijing (2012), 'China's photovoltaic industry on brink of bankruptcy: report', 8 August 2012, accessed 1 April 2016 at http://english.caijing.com.cn/2012-08-08/111998243.html.

Chen, G. (2012), *China's Climate Policy*, London: Routledge.

Chen, G. (2015), 'From mercantile strategy to domestic demand stimulation: changes in China's solar PV subsidies', *Asia Pacific Business Review*, 21 (1), 96–112.

China Business Times (2013), 'Different impact of Suntech's default in domestic and overseas bond market' [*Shangde jiaocha weiyue jingneiwai zhaiquan youbie*], 28

March 2013, accessed 1 April 2016 at http://business.sohu.com/20130328/n370658622.shtml.

China's 12th Five-Year Plan on Energy Development [in Chinese] (2013), accessed 7 December 2016 at http://www.gov.cn/zwgk/2013-01/23/content_2318554.htm.

China's 13th Five-Year Plan on Energy Development [*kezaisheng nengyuan fazhan shisanwu guihua*] (2016), accessed 8 May 2017 at http://www.ndrc.gov.cn/zcfb/zcfbtz/201612/W020161216659579206185.pdf.

China's State Electricity Regulatory Commission (2011), *Power Industry Monitoring and Supervision Report 2010* [*2010 niandu dianjia zhixing ji dianfei jiesuan qingkuang tongbao*], Beijing: State Electricity Regulatory Commission.

Ecoworld (2008), 'Wind power in China', accessed 3 December 2015 at http://www.ecoworld.com/energy-fuels/wind-power-in-china.html.

EU ProSun (2013), 'European Commission confirms massive illegal subsidies to Chinese solar manufacturers', 28 August 2013, press release, Brussels, accessed 20 August 2018 at https://www.eureporter.co/economy/2013/08/29/european-commission-confirms-massive-illegal-subsidies-to-chinese-solar-manufacturers/.

European Commission (2012), 'News archive: EU initiates anti-dumping investigation on solar panel imports from China', accessed 1 July 2017 at http://trade.ec.europa.eu/doclib/press/index.cfm?id=829.

Forbes (2009), 'Weaknesses in Chinese wind power', 20 July 2009, accessed 9 January 2017 at www.forbes.com/2009/07/20/china-wind-power-business-energy-china.html.

Haley, U. C. V. and G. T. Haley (2013), *Subsidies to Chinese Industry: State Capitalism, Business Strategy, and Trade Policy*, New York: Oxford University Press.

Li, J. and L. Ma (2009), 'Background paper: Chinese renewables status report', accessed 20 December 2016 at http://ren21.net/Portals/0/documents/Resources/Background_Paper_Chinese_Renewables_Status_Report_2009.pdf.

Li, L. (2010), 'Brief analysis of current China's PV industry: part II' [*woguo guangfuchanye fazhanxianzhuang jianxi zhier*], *Solar & Renewable Energy Sources* [*yangguang nengyuan*], 2, 36–9.

Meiri Jingji Xinwen (2012), 'Reflection on PV expansion: impulse from local governments' [*guangfu yuejin zhi hangye fansi: difangzhengfu de chongdong yu niai*], 22 June 2012, accessed 5 April 2016 at http://finance.sina.com.cn/chanjing/cyxw/20120622/014712378110.shtml.

PEW Charitable Trusts (2010), *Who's Winning the Clean Energy Race? Growth, Competition and Opportunity in the World's Largest Economies*, accessed 2 March 2017 at www.pewtrusts.org/uploadedFiles/wwwpewtrustsorg/Reports/Global_warming/G-20%20Report.pdf?n=5939.

PV Magazine (2018), 'Trump levies graduated tariffs starting at 30%, exempts 2.5 GW in cells', accessed 2 March 2018 at https://www.pv-magazine.com/2018/01/23/trump-levies-graduated-tariffs-starting-at-30-exempts-2-5-gw-in-cells/.

Renewable Energy Policy Network for the 21st Century (REN21) (2009), report, *Recommendations for Improving the Effectiveness of Renewable Energy Policies in China*, accessed 6 January 2015 at www.ren21.net/pdf/Recommendations_for_RE_Policies_in_China.pdf.

Renewable Energy Policy Network for the 21st Century (REN21) (2013), *Renewables 2013 – Global Status Report*, accessed 9 July 2016 at http://www.ren21.net/Portals/0/documents/Resources/GSR/2013/GSR2013_lowres.pdf.

Reuters (2010), 'U.S. challenges China wind power aid at WTO', 23 December

2010, accessed 1 February 2017 at https://www.reuters.com/article/us-usa-china-windpower/u-s-challenges-china-wind-power-aid-at-wto-idUSTRE6BL3EU20101222.

Reuters (2011), 'China agrees to halt subsidies to wind power firms', 7 June 2011, accessed 2 February 2017 at https://www.reuters.com/article/us-china-windpower/china-agrees-to-halt-subsidies-to-wind-power-firms-idUSTRE7561B920110607.

Salim, R.A. and S. Rafiq (2012), 'Why do some emerging economies proactively accelerate the adoption of renewable energy?', *Energy Economics*, 34 (4), 1051–7.

Shilling, A. G. (2012), 'India's accelerating shift toward free markets', *Bloomberg View*, accessed 26 December 2014 at http://www.bloombergview.com/articles/2012-12-17/india-s-accelerating-shift-toward-free-markets.

Sina.com News (2011), 'Abolishment of wind power subsidy confirmed', [*fengdian butie queding beiquxiao*], accessed 5 January 2017 at http://finance.sina.com.cn/chanjing/cyxw/20110618/141710012325.shtml.

US Department of Commerce (2012), Fact Sheet, 'Commerce finds dumping and subsidization of crystalline silicon photovoltaic cells, whether or not assembled into modules from the People's Republic of China', accessed 2 December 2017 at https://enforcement.trade.gov/download/factsheets/factsheet_prc-solar-cells-ad-cvd-finals-20121010.pdf.

US Trade Representative (2018), Fact Sheet, 'Section 201 cases: imported large residential washing machines and imported solar cells and modules', accessed 24 January 2018 at https://ustr.gov/sites/default/files/files/Press/fs/201%20Cases%20Fact%20Sheet.pdf.

Yang, M. and R. Pan (2010a), 'China's wind power industry: from infant to growth stage', Background Brief No. 525, Singapore: East Asian Institute.

Yang, M. and R. Pan (2010b), 'Harvesting sunlight: solar photovoltaic industry in China', Background Brief No. 562, Singapore: East Asian Institute.

3. Central authorities' top-down approach of promoting renewable energy

China's 11th, 12th, and 13th Five-Year Plans (2006–10; 2011–15; 2016–20), the Renewable Energy Law (2005) and its amendment (2009), as well as the Medium- and Long-Term Development Plan for Renewable Energy (2007) have provided programmatic guidance systems for the government's support of renewable power generation and related manufacture. The country institutionalized its efforts to develop clean energy capacity by passing the pioneering Renewable Energy Law in 2005, which defined renewable energy as non-fossil energy, which includes wind energy, solar energy, water energy, biomass energy, geothermal energy, and ocean energy (PRC Renewable Energy Law 2005, Article 2).

The law prompted the government to utilize financial subsidies and tax incentives for the development of renewable sectors. A PEW Charitable Trusts (2010, p. 15) report found that as early as 2009, China's clean energy investments soared more than 50 percent to US$34.6 billion, taking the lead among G-20 nations for the first time and pushing the United States into second place. Nevertheless, many of the central government's supportive policies in the beginning proved to be unrealistic and unattainable, only starting to take effect when market conditions started to become more mature and interest groups more cooperative. China's centralized efforts in renewable energy development were hampered by legal, congressional, and party leadership constraints embedded in the fragmented and pluralistic bureaucratic politics.

FROM SUBSIDIZING EQUIPMENT MANUFACTURING TO STIMULATING RENEWABLE POWER GENERATION

China's renewable energy policy that had prioritized renewable equipment manufacturing, export revenues, and employment increase led to serious imbalance between a burgeoning manufacturing industry and weak demand in an underdeveloped domestic market of renewable energy generation. *The Economist* (Wooldridge 2012, p. 13) in a special issue linked China's great leap forward in solar and wind power with a formidable state capitalism model that can enjoy some success in tackling second-generation infrastructure problems and mandating higher environmental standards. Under the framework of the Renewable Energy Law and national green targets, detailed institutional incentives and disincentives have been designed by various layers of the government to promote renewable energy development in China.

In 2008, in response to the global financial crisis, the central government gave research and development (R&D) subsidies totaling 4 billion yuan (US$585 million) for the development and industrialization of core renewable energy equipment, including key components for wind turbines and advanced silicon technologies for solar photovoltaics (PV), with another 2 billion yuan allocated for R&D in biofuels (Renewable Energy Policy Network for the 21st Century 2009, pp. 20–21). Yet even at that time, astute researchers (Li 2010; Yang and Pan 2010) had forewarned about the potential overcapacity problems as a result of over-subsidy and overinvestment. Li cautioned about China's overreliance on external demand when boosting its solar panel manufacture in the infant stage, categorizing the asymmetrical problem as Both-Ends-Out (*liangtou zaiwai*), which implied that China imported 95 percent of its PV raw materials and exported 95 percent of its finished products due to an underdeveloped domestic market (see Chapter 2). The topic caught the attention of a research team led by Zhang and Andrews-Speed, who attributed China's wind and solar overcapacity and overdependence on foreign markets to the lack of appropriate interactions between the renewable energy policy and the renewable energy industrial (manufacturing) policy (Zhang et al. 2013).

While contributing the largest portion to global PV production, China accounted for a disproportionately low share of worldwide accumulative PV power generation capacity. Taking advantage of the global trend focusing on fossil fuel replacement amidst growing public awareness of climate change, China's early support for the PV industry focused on reshaping factors of production rather than on home demand conditions. Dominated by concerns that the exorbitant cost of installing and operating PV facilities might become a heavy financial burden on the government, the state apparatus at both the central and local levels preferred to generously subsidize PV production aimed at the overseas market rather than endorsing nationwide feed-in tariffs that supported domestic PV consumption until 2011. This strategy under the Chinese model of state capitalism helped the country to net bountiful trade surplus while its solar panel exports enjoyed twofold benefits from domestic production subsidies and overseas consumption-end subsidies in the name of carbon dioxide mitigation. Ironically, the stimulation of PV production in the name of green development and sustainable growth did not contribute much to the improvement of China's own air quality or climate change mitigation causes; on the contrary, China has ended up with serious pollution, high energy consumption and extensive debt problems associated with domestic overcapacity. Only under these circumstances did the Chinese government shift its state strategy from subsidizing PV producers to subsidizing demand-end power generation, extending an above-market-price feed-in tariff across the entire country for the first time in 2011.

Nevertheless, the various state subsidies offered to domestic PV producers have not ceased under the mercantile model of state capitalism, which still values the expansion of trade surplus and accumulation of foreign reserves as its long-term national goals. The government has bailed out major solar panel makers like Wuxi Suntech and Shanghai Chaori from time to time. Shanghai Chaori, which defaulted in the country's onshore bond market, received a guarantee of as much as 788 million yuan (US$128.4 million) from the China Great Wall Asset Management Corporation on the solar panel maker's 1 billion yuan of defaulted notes (*Bloomberg News* 2014). China Great Wall Asset Management Corporation, established by the Chinese government in 1999, is a state-backed fund that helps large-sized companies, most of which are state-owned enterprises, to restructure their bad debts in the domestic market. Despite its emergence as a major producer of renewable energy equipment, China's domestic demand of these products lagged behind that of some industrialized countries for quite a long time. Unlike its persistent interest in hydro and thermal power generation for the sake of energy security, China turned its attention to wind and solar power generation much later, largely in response to the new context of climate change and redundant manufacture capacity of renewable equipment.

To understand the powerful role the Chinese government subsidies have played in influencing business strategies and national competitive advantages in the renewable sector, one has to revisit Porter's classical description of government's indirect influence on national competitive advantages through the 'diamond' of four determinants, namely: factor conditions; firm strategy, structure, and rivalry; demand conditions; and related and supporting industries (Porter 1998, pp. 72–128). Porter's theory links government industrial policies to the economic successes of some East Asian countries, whose comparative advantages benefit indirectly from these policies that reshape the circumstances of the four determinants of national advantages in a positive way (Porter 1998, p. 126). In these developmental states, factor conditions such as labor, land, raw materials, capital, and infrastructure are often remolded by the government through subsidies and supportive policies, while indigenous demand can also be reframed by local product standards, government procurement, and consumption subsidies.

Under the Chinese model of state capitalism, in which many manufacturing industries are backed by proactive government policies, China's wind turbine and solar panel production has long featured the highest localization rates for its equipment, the lowest technological threshold, the largest number of small- and medium-sized businesses, and the greatest expansion in the global PV manufacturing industry. The domestic manufacturing boom, or the development of supporting industries from Porter's perspective, has justified the Chinese government's approach to promoting inland wind- and solar-generated power, or the improvement of demand conditions in accordance with Porter's theory. In another word, the once-expensive renewable power generation gradually became cost-competitive vis-à-vis conventional power and thus needed fewer subsidies on a per kW-hour basis. China's wind turbine industry has witnessed rapid expansion since 2008, bringing down turbine prices by as much as 20–25 percent in Western markets and more than 35 percent in China between 2008 and 2012 (Renewable Energy Policy Network for the 21st Century 2013, p. 54). The costs of operating and maintaining wind farms dropped significantly due to increased competition among contractors and improved turbine performance.

China's solar panel manufacturing has been facing a far worse overcapacity problem as a result of governmental over-subsidy and overinvestment. Since the early 2000s, the center of gravity for solar PV production has shifted from the United States to Japan, to Europe, and to Asia, especially China, which quadrupled its polysilicon solar panel manufacturing capacity between 2009 and 2011. Within just four years, between 2009 and 2012, the international price of solar panels was cut by more than 75 percent largely due to an aggressive capacity build-up in China. To alleviate the overcapacity problem, the Chinese government has therefore redirected its energy subsidies to support PV power generation in order to grow the domestic renewable power market substantially and absorb the oversupply that faced overseas boycott.

CHINA'S RENEWABLE ENERGY LAW: STATUTORY AND ENFORCEMENT FLAWS

In its early stage of promoting renewable energy generation, the central Chinese government relied on the ad hoc legal and policy frameworks constituted by the PRC Renewable Energy Law and various national plans on energy development. Nevertheless, such a centralized and top-down approach to fostering renewable energy development through laws, plans, and regulations has innate institutional flaws in terms of enforcement at both central and local levels. Similar to the many Chinese laws and regulations written in a rush, the PRC Renewable Energy Law was also improvised in the global political context of climate change and hence used quite vague and imprecise language by which the law can be easily bypassed. Both the law and its amendment in 2009 are more like policy statements rather than laws.

The PRC Renewable Energy Law was enacted by the National People's Congress (NPC) in 2005, the same year when the milestone Kyoto Protocol took effect globally. It stipulated that the 'energy administration department under the State Council exercises unified control over the exploitation of renewable energy nationwide' (PRC Renewable Energy Law 2005, Article 5), but it did not provide legal details to ensure that such a pivotal role was played by the energy administration department, which was affiliated to the National Development and Reform Commission (NDRC) at that time, in promoting renewable energy development. Confusingly, the law also stated that 'relevant departments under the State Council are responsible for administration of the work related to the exploitation of renewable energy within the limits of their respective duties' (PRC Renewable Energy Law 2005, Article 5), which diluted the power of the energy administration department in this policy area. 'Relevant departments' may refer to the many ministries and departments under the State Council, China's chief executive organ, which are entitled to enact administrative regulations and rules on specific energy or environmental issues.

The PRC Renewable Energy Law made a similar regulatory glitch as regards local enforcement. Not only did Article 5 stipulate that 'the energy administration departments of the local people's governments at or above the county level are responsible for the administration of exploitation of renewable energy within their respective administrative areas', but it also authorized the 'relevant departments' of the local people's governments at or above the county level to be responsible for administration of the work related to the exploitation of renewable energy within the limits of their respective duties.

At both central and local levels, the law failed to offer clear-cut regulations regarding the boundaries of authority between energy departments and other 'relevant departments' over renewable energy promotion. In addition to the NDRC and its local branches, the Ministry of Environmental Protection, the Ministry of Land Resources, the Ministry of Industry and Information Technology, the Ministry of Science and Technology, the Ministry of Agriculture, and state-owned grid companies, as well as their local subsidiaries, are all authorized to participate in renewable energy regulation. The Ministry of Industry and Information Technology, for example, is authorized by the law to 'set and publish the technical standards of the State for grid-connected power generation with renewable energy and other standards of the State for the technology and products related to renewable energy, for which the technical requirements need to be uniform throughout the country' (PRC Renewable Energy Law 2005, Article 11).

Nevertheless, with regard to those technical requirements which are not covered by the standards of the State as mentioned in the preceding paragraph, the Law also allows other 'relevant departments' under the State Council to establish relevant industrial standards, which shall be submitted to the Ministry of Industry and Information Technology for the record (PRC Renewable Energy Law 2005, Article 11). This actually dilutes the Ministry of Industry and Information Technology's authority in setting industrial standards for renewable sectors. As the Renewable Energy Law provides only a nationwide legal framework, local courts and governmental departments across China also have the right to publicize detailed local regulations for implementing the PRC Renewable Energy Law according to special local conditions. All these central and local laws, regulations and rules constitute China's renewable energy legal system in a broad sense.

Typically, actions are encouraged but rarely required and even where concrete duties are stated, only little guidance is provided on procedures and specific goals (Beyer 2006, p. 205). Considering the comparable disadvantage of generating electricity from renewable sources, the Renewable Energy Law addressed the core issues of pricing and fee-sharing for on-grid renewable energy through a government-set or government-guided pricing system. Article 14 of the Law on Renewable Energy, for example, requires a power grid enterprise to:

> sign a grid-connection agreement with the enterprise of power generation with renewable energy that has legally obtained an administrative license or has submitted the project to be constructed for the record, and buy the entire quantity of the grid-connected power generated with renewable energy within the coverage of their power grids, and provide grid-connection services for the generation of power with renewable energy.

The law, however, does not specify the punitive measures for grid companies if they refuse to do so. After only four years of passing the Renewable Energy Law, the government found that, although the country had rapidly increased its installed capacity of renewable energy, much of this capacity was not promptly connected to the grid and that not all the power being generated was being purchased as required by the law.

When the Renewable Energy Law was amended in 2009, the top legislature required electricity grid companies to buy all the power produced by renewable energy generators and authorized the State Council energy department, in conjunction with the state power regulatory agency and the State Council finance departments, to determine of the quantity of electricity generated from renewable energies as percentage of the total quantity of electricity generated during the planned period (PRC Amended Law on Renewable Energy 2009, Article 14), a regulation system similar to the Renewable Portfolio Standards (RPS) adopted by some US states. The amendment made it clear that power enterprises refusing to buy power produced by renewable energy generators would be fined up to an amount double that of the economic loss of the renewable energy company (PRC Amended Law on Renewable Energy 2009, Article 29). Article 15 of the PRC Renewable Energy Law 2005 says 'the State supports construction of independent system of power generated with renewable energy in areas not covered by power grids,' but it does not give details as to the provision of this state support to localities. Article 25 states 'financial institutions may offer preferential loans with financial interest subsidy to projects for exploitation of renewable energy that are listed in the national development guidance catalogue of the renewable energy industry and meet the requirements for granting loans,' but the regulation does not specify the punitive measures if financial institutions refuse to do so, or how they would be compensated by the state if such financial support results in bad debts.

When designing the legal system on renewable energy, Chinese lawmakers expected local governments to play a cooperative role in enforcing the Renewable Energy Law. In practice, however, this high degree of administrative cohesion does not exist, and the law failed to anticipate the possibility that certain government interests might diverge sharply from those of the central departments and create a major obstacle to strict enforcement of both national and local legislation (Beyer 2006, p. 207). As stakeholders in local fossil fuel companies, local authorities in many cases tend to help these companies ignore environmental standards and regulations that may hinder investment inflow and economic growth, or hamper the development of their low-carbon rivals.

For the allocation of group resources between congressional and executive liaison from central to local levels, congressional lobbying was the first priority of most interest groups because people's congresses dominated the renewable energy legitimation process, where net marginal benefits were maximized, and played a consultative role in the policy formulation process. The executive bureaucracy in the government received secondary attention in the beginning, but gradually it became the focal point for lobbying activities in the follow-up regulation formulation and policy implementation processes. After the Renewable Energy Law was enacted by the National People's Congress, the State Council departments formulated a number of regulations for the implementation of the law, including the Regulation on Renewable Energy Power Generation (2006), the Trial Regulation on the Fee and On-grid Tariff Sharing of Renewable Energy Power Generation (2006), the Trial Regulation on the Management of Additional Grants to Renewable Energy Power Generation (2006), and Supervisory Regulation on Grid Companies' Purchase of Electricity Generated from Renewable Power (2007) (Ma 2016). Those regulations relating to on-grid power tariff were promulgated by the NDRC, while those pertaining to subsidization and approval of renewable energy projects were drafted by the Ministry of Finance or the Ministry of Construction, predecessor of the Ministry of Housing and Urban-Rural Development. (For interactions between energy interest groups and administrative departments, see Figure 8.1.) If interest groups are to accomplish their basic goals, they have to be attentive to these regulations drafted by administrative apparatus, plans from the Ministry of Finance and its local branches, which must appropriate funds and subsidies for supportive programs, as well as the aspirations of other government authorities that undertake these programs and supervise policy implementation.

The legitimation of the PRC Renewable Energy Law and its amendment represented Chinese law-makers' goodwill towards developing renewable energy and relevant equipment manufacturing in the global context of climate change, but bureaucratic attentiveness to congressional wishes should not imply that the bureaucracy is powerless relative to legislative demands. It is difficult for the NPC to shake off the label of a 'rubber stamp' given that a full plenary session has always deferred to decisions already made by the ruling party and governments at various levels. In theory, the administrative procedures of every executive agency are subject to fiscal and procedural oversight by the NPC. In practice, however, the legislative branch has difficulty policing the executive branch due to enormous congressional constraints in the authoritarian party-state. Although some may perceive the NPC's assertiveness on specific issues as an indication of greater autonomy, evidence suggests that the occasional deviations from rubber-stamp unanimity should be more accurately seen as manifestations of intra-Party disagreement or rivalry among different interest groups.

CONGRESSIONAL AND PARTY LEADERSHIP CONSTRAINTS IN BUREAUCRATIC POLITICS

The political environment of the administrative agency affects the agency's mode of interaction with interest groups through congressional, leadership, and judicial constraints. Differences in these constraints should help explain why different executive agencies establish differing relationships with their interest group constituencies (Chubb 1983, p. 46). China's unique one-party political system predetermines the quality of the country's governance of renewable energy promotion and other sustainable development issues. Concerning the interrelations among the ruling party, the legislature (NPC), and the cabinet (State Council), the Party plays a dominant role in all decision-making processes as the legislature and the cabinet include Party members, and its ruling status is stipulated by the Constitution.

It is efficient to enforce central policies in various parts of the vast territory at various levels under such a unitary political system, but this causes intractable governance problems such as lack of supervision, corruption, and loose law enforcement. In terms of environmental and energy governance, the growing attention paid by the top Party leaders at the central level to pollution and other ecological problems has mobilized the whole governmental system to make more effort in the green area; however, the monolithic power structure from top down to the grassroots in a multilayered structure prevents such problem from being effectively solved when local party bosses are not incentivized to push forward green causes.

Although all three systems (the Party, the administration, and the legislature) are led by Party members at the various levels, there is still an implicit pecking order of their power in daily political life. The wishes of Party secretaries of the Communist Party of China (CPC) committees at all five governmental levels (central, provincial, municipal, county, township) dominate in the affairs within the realm, while the administration chiefs at various layers, who are responsible for routine and detailed administrative work, have to listen to the Party committee's voice during decision-making on important or controversial issues. The system of People's Congress (legislature) at all levels is the least powerful branch of the three. The NPC, labeled by Western media as the 'rubber stamp' is regarded by the Constitution as the nation's top legislative body, enacting both basic environmental/energy laws and special environmental/energy laws. In practice, however, the State Council (administration) drafts proposed laws and refers them to the NPC and its Standing Committee. The State Council plays this influential legislative role because it sits at the top of a vast bureaucracy that includes virtually all ministries, commissions, and other administrative agencies, and thus it has considerable access to expertise and resources (Ma and Ortolano 2000, p. 14).

Meanwhile, local law-makers also have the right to promulgate their own statutes in compliance with the principles set by the PRC Renewable Energy Law. The provinces of Hunan, Shandong, and Heilongjiang enacted their decrees on rural renewable energy development in 2005, 2007, and 2008, respectively, but many other provinces have failed to promulgate similar local regulations to facilitate the implementation of the Renewable Energy Law (Li and Yue 2010, p. 8). As compared with the members of the NPC special committees, the expertise and enthusiasm of local legislators decline down the ranks from province to county level. The NPC at various levels usually holds plenary sessions once a year, and delegates have limited power in supervising law enforcement in practice. China lacks an independent judicial system, with courts at all levels under the absolute Party leadership. Under such systems, the Party committee and the administration have the upper hand while the legislature and courts are somewhat marginalized in the game of power and thus lack the authority to enact checks and balances. Without sufficient supervision, selective law enforcement can become widespread, and *guanxi*, or social connections, are essential for energy producers to get subsidies or grants from administrative departments such as local bureaus of finance or construction. If entrepreneurs maintain good relations (*guanxi*) with local officials through bribery in forms of cash, valuable gifts, invitations to banquets and all-expense-paid trips abroad, it is not difficult for them to avoid complying with the strict environmental standards and thus lower production costs.

Top CPC leaders, stimulated by global action against climate change, urged Chinese officials and law-makers to enact the Renewable Energy Law and relevant regulations with the implementation of the Kyoto Protocol on the horizon. The Committee on Natural Resources and Environmental Protection under the NPC was supported by the NDRC and other ministries in the State Council in its efforts to formulate the Renewable Energy Law. Nevertheless, such a national law, which was drafted in a hasty way, includes mainly principles and guidelines and is therefore not practicable for local authorities to implement (Niu and Zhang 2008, p. 24).

Due to the lack of expertise in specific categories of renewable energy in the NPC Committee on Natural Resources and Environmental Protection, the law tries to promote all kinds of renewable energies in a holistic manner instead of addressing the distinguishing features of different renewables (Li and Yue 2010, p. 8). The law in many places simply duplicates mature schemes such as 'cost-sharing', 'feed-in tariffs', 'renewable portfolio targets', and 'mandatory grid-connection' from similar statutes in European nations and the United States (Chen 2012, p. 47), without taking China's indigenous economic and political conditions into consideration. Even those supplementary regulations from various administrative departments are fragmented and sometimes conflict with one another, leading to poor coordination, enforcement, and supervision (Niu and Zhang 2008, p. 24). The ineffective oversight from the subcommittees of the NPC, especially the one on Natural Resources and Environmental Protection, has failed to check interest groups' influences on the bureaucracy, resulting in miscellaneous regulations from various ministries and commissions in the State Council.

The State Council, China's chief administrative authority chaired by the Premier, consists of over 80 ministries, administrations, and bureaus, and is in charge of overall socioeconomic policy, making its supervisory role too vague and broadly defined for regulating energy producers or equipment manufacturers across the country. Most of its policy decisions simply are of too little political consequence to command top party-state leaders' attention. Due to the sheer size of the bureaucracy and administrative rules that regulate inter-agency procedures, the influence of the senior-most leaders may have difficulty penetrating below the upper layer of state apparatus. With the establishment of six permanent special committees in 1983, the NPC has been responsible for drafting and formalizing a noteworthy corpus of laws, thereby contributing to the making of China's legislative system, which was virtually non-existent under Mao (Lieberthal 2004, p. 176). Yet the lack of expertise and representativeness of the NPC delegates prevents them from producing high-standard, pertinent laws promoting renewable energy and ecological protection, making them vulnerable to lobbying activities from powerful interest groups, either within or outside the executive bureaucracy. These congressional and leadership constraints in bureaucratic politics are the key institutional factor that often skews the central leadership's priorities in the process of policy formulation, legitimation, and implementation.

CENTRAL LEADERSHIP'S CHANGES TO POLICY PRIORITIES AND RESISTANCES FROM INTEREST GROUPS

In each stage of its renewable sector development, changes to Chinese central leadership policy priorities have always been triggered by both international and domestic pressures. In the 1980s and 1990s, China's response to global atmospheric problems, including ozone depletion and climate change, were overshadowed by such intransigent propositions in the realm of high politics as supremacy of national sovereignty, development-first for the Third World, and the industrialized–developing nation dichotomy, but its recalcitrant negotiating positions were gradually softened by the external impact of international actors on China's domestic political process and foreign policy-making.

When the milestone Kyoto Protocol was about to take effect in the early 2000s, China, as then the world's second-largest carbon emitter, felt growing pressure on its mitigation actions and energy policy. The statutory and enforcement flaws manifested in the Renewable Energy Law resulted from such improvised changes to policy priorities by China's top leadership, unpreparedness of central and local enforcement apparatus, and resistances from within the bureaucracy and various interest groups. Inspired by Germany's 2000 Renewable Energy Sources Act, China enacted the Renewable Energy Law in 2005 to mark its leaders' fresh and firm commitment to the development of low-carbon energy alternatives in the context of global climate consciousness. Yet this policy change, from prioritization of fossil fuels to prioritization of renewable energy, came so abruptly and bluntly that most of its legislative and executive departments as well as relevant interest groups were caught unprepared. The central leadership tried to prompt government departments at various levels to provide financial subsidies and grants for local renewable energy producers through new schemes originating in the Western context, including 'feed-in tariffs', 'mandatory grid-connection', 'cost-sharing', and 'renewable portfolio target'. The goal of feed-in tariffs, which are usually higher than benchmark tariff levels, is to offer cost-based compensation to renewable energy producers, providing price certainty and long-term contracts that help finance renewable energy investments (Couture and Gagnon 2010, p. 955).

Due to fiscal constraints and vested interests in conventional energy sectors, the 'feed-in tariffs' for wind power generation (see Chapter 6) were perceived by many local grids and governments as being too high and unrealistic for execution. The feed-in tariffs for wind electricity ranged between 0.51 yuan (US$0.078)/kWh and 0.61 yuan/kWh in 2009 (see Chapter 6), while the benchmark tariffs for coal-fired electricity in those localities fluctuated around 0.3 yuan/kWh. The law forced grid companies to buy electricity generated by local windmills, usually at the price level of coal-fired tariffs, while the gap between that and the feed-in tariff level had to be compensated by a nationwide electricity price surcharge in the name of promoting renewable energy (Yu 2015, p. 123). This surcharge, however, was never sufficient to offset the price gap, especially in the context of sharp increase in the number of wind turbines and surging electricity generated from them. Despite the NDRC's effort to increase the renewable energy surcharge from 0.001 yuan/kWh in 2006 to 0.004 yuan/kWh in 2009, and then to 0.008 yuan/kWh in 2011, the deficit between the total surcharge collected across the country and the amount of subsidy offsetting the benchmark coal-fired electricity tariffs and feed-in tariffs was 1.3 billion yuan in 2009, quickly rising to 10 billion yuan in 2011 and over 20 billion yuan in 2012 (Yu 2015, p. 123). For a long time, due to budgetary constraints, the NDRC, the State Electricity Regulatory Commission, and the Ministry of Finance had suspended the nationwide allocation of subsidies used to support wind power feed-in tariffs. The state-owned grid companies in many cases were either unable or unwilling to get newly installed wind or solar PV projects to be connected to grids, thanks to their remote locations and lack of sufficient fiscal support.

In the first few years after the enactment of the Renewable Energy Law, there was a sharp increase in wind power generation projects, thanks to promises of support from both central and local governments and the enthusiasm of local interest groups benefiting from infrastructure construction projects. The number of solar power generation projects, however, remained at an insignificant level due to the high cost of solar panels at that time and insufficient fiscal support from the Ministry of Finance. In 2006, the cost of electricity generated from solar power was about 3 yuan/kWh, while that of a typical coal-fired power plant was only around 0.22 yuan/kWh, and that of a wind power plant averaged at about 0.6 yuan/kWh. At that time, many newly installed solar PV power projects were not connected to the grid due to their remote locations or resistance from state-owned grid companies with strong ties to conventional thermal or hydropower companies. There was no unified on-grid tariff for solar PV until 2011, when the government formally announced a feed-in tariff of 1 yuan/kWh for solar PV projects scheduled to be completed after 2011 and another tariff of 1.15 yuan/kWh for projects completed before the end of 2011 (National Development and Reform Commission 2011). As the solar PV feed-in tariffs were even higher than those for wind power generation, both central and local governments were actually facing more stringent fiscal challenges in subsidizing solar PV power generation.

The outbreak of the global financial crisis in 2008, coupled with slumping external demand for Chinese products, forced China's top leaders to temporarily shift their policy priorities from boosting renewable energy generation to production of renewable energy equipment such as wind turbines and solar panels. As compared with its nascent effort to increase renewable energy generation, this round of policy shift prioritizing mercantile targets was forcefully implemented by executive departments, and widely supported by local interest groups that had huge stakes in export-oriented manufacturing sectors.

China's explosive development in the new field of wind turbine and solar PV manufacturing, which cannot be explained from the conventional perspective of comparative advantages, should instead be scrutinized through the theoretical lens of state capitalism, a common East Asian developmental model in which states play a significant and sometimes decisive role in industrialization and markets (Johnson 1995; Hobson 1997; Fligstein and Zhang 2011). As with other manufacturing sectors, the Chinese government, at both central and local levels, provided local manufacturers along the industrial chains with various forms of subsidies, which included free or low-cost loans, tax rebates, research grants, cheap land, energy subsidies, and technological, infrastructure, and personnel support. With the efficient enforcement by the bureaucracy and cooperation from interest groups, the policy received huge paybacks as China's manufacturing sector for producing wind turbines, solar panels, and modules boomed and became the new engine of economic growth during the aftermath of the financial crisis. However, this also led to serious trade disputes with the United States and the European Union, both of which imposed punitive tariffs on Chinese exports of renewable energy equipment (see Chapter 2).

Besides external pressures, this manufacture-oriented policy also led to severe imbalance between the burgeoning manufacturing industry and weak demand in the underdeveloped domestic market of renewable energy generation (see Chapter 2). Faced by the fresh imperatives from trade spats, post-Kyoto Protocol climate discourses, and domestic overcapacity, the Chinese leadership had to shift policy focus again to the acceleration of domestic renewable power generation, with the aim of absorbing redundant equipment production and increasing the share of low-carbon alternatives in its energy mix. In retrospect, many of the central government's supportive policies for renewable energy production proved to be unrealistic and unattainable in the beginning, only starting to take effect when market conditions became more mature, and administrative departments and interest groups became more cooperative.

China's Five-Year Plans have been focusing on targets of newly installed renewable power generation capacity (e.g., 100 GW for newly added wind power capacity by 2015 and 200 GW by 2020), but they do not pay equal attention to the efficiency of using these newly added capacities in various renewable sectors, leading to huge wastage and overcapacity in the regions. As the frontrunner among China's various renewable energy sectors, wind power generation has experienced enormous infrastructure (lack of grid-connection services) and fiscal (short of subsidization) bottlenecks, which have prevented tapping the full potential of windmills. If full use had been made of the untapped wind power capacity of windmills, a total of 20 TWh and 16.2 TWh of electricity would have been generated in 2012 and 2013, respectively (Chinese Renewable Energy Industries Association (CREIA), Chinese Wind Energy Association (CWEA) and Global Wind Energy Council (GWEC) 2014, p. 16).

China's model of state capitalism, which has been extensively and prominently discussed to understand the country's meteoric economic growth since the early 1980s (Wong 2004, pp. 345–62), has played a crucial role in reshaping the domestic industrial and market conditions concerning renewable energy sectors. In the face of a downturn in demand of external renewable equipment and severe internal overcapacity, China's state capitalism has been forced to evolve from a mercantile stage that entailed providing more subsidies to help domestic manufacturers of wind turbines and solar panels, to a new stage of domestic consumption involving the introduction of policies to stimulate indigenous renewable power generation to absorb redundant capacity.

The Renewable Energy Law was amended in 2009 to make it mandatory for electricity grid companies to buy all the power produced by renewable energy generators, and to authorize the State Council to determine the percentage share of electricity generated from renewable energies in the total quantity of electricity (PRC Amended Law on Renewable Energy 2009, Article 14). Nevertheless, as China's power industry is still highly regulated by the state, the feed-in tariffs, renewable portfolio targets, and ambitious low-carbon plans have seldom reflected market fundamentals or been realistic, leading to huge wastage and unsustainable expansion in the renewable energy sector. The government-set or government-guided schemes of feed-in tariffs, cost-sharing, and renewable portfolio targets were not well received by various stakeholders at both central and local levels for a long time, resulting in serious market distortion in the country's nascent renewable energy industry. The on-grid electricity tariffs, for instance, are still set by government regulators and play a pivotal role in the profit margins and even survival of various power generators.

Compared with conventional power corporations which are state-owned behemoths subject to state plans and government edicts, wind and solar power plants are mostly small and medium-sized generators with diverse ownership, and are therefore more sensitive to fluctuations in power pricing, production costs, and demand conditions. Government plans, which are more precise in predicting future development trajectories of conventional energies, often lag behind in forecasting the proliferation rate of wind and solar power plants, which face higher supply and demand elasticity due to government subsidies, feed-in tariffs, and production costs. The Chinese leadership's centralized, top-down approach has its limits in promoting renewable energy production in such a volatile market environment.

REFERENCES

Beyer, S. (2006), 'Environmental law and policy in the People's Republic of China', *China Journal of International Law,* 5 (1), 185–211.

Bloomberg News (2014), 'China's first bond defaulter gets guarantee from bad bank', 8 October 2014, accessed 5 January 2017 at http://www.bloomberg.com/news/articles/2014-10-07/chaori-solar-restructuring-plan-seeks-haircuts-up-to-80-.

Chen, G. (2012), *China's Climate Policy*, London: Routledge.

Chinese Renewable Energy Industries Association (CREIA), Chinese Wind Energy Association (CWEA) and Global Wind Energy Council (GWEC) (2014), 'Problems and prospect for China's wind power generation industry' [*zhongguo fengdian chanye fazhan wenti ji zhanwang*], *China Electrical Equipment Industry* [*dianqi gongye*], 11, 15–18.

Chubb, J. E. (1983), *Interest Groups and the Bureaucracy: The Politics of Energy*, Stanford: Stanford University Press.

Couture, T. and Y. Gagnon (2010), 'An analysis of feed-in tariff remuneration models: implications for renewable energy investment', *Energy Policy*, 38 (2), 955–65.

Fligstein, N. and J. Zhang (2011), 'A new agenda for research on the trajectory of Chinese capitalism', *Management and Organization Review*, 7 (1), 39–62.

Hobson, J. M. (1997), *The Wealth of States*, Cambridge: Cambridge University Press.

Johnson, C. (1995), *Japan: Who Governs? The Rise of the Development State*, New York: Norton.

Li, L. (2010), 'Brief analysis of current China's PV industry: part II' [*woguo guangfuchanye fazhanxianzhuang jianxi zhier*], *Solar & Renewable Energy Sources* [*yangguang nengyuan*], 2, 36–9.

Li, Y. and X. Yue (2010), 'Construction of China's renewable energy law system' [*unwoguo kezaishengnengyuan falu tixi de goujian*], *Gansu Social Sciences*, 2, 7–11.

Lieberthal, K. (2004), *Governing China: From Revolution Through Reform*, New York: W. W. Norton.

Ma, J. (2016), 'Tenth anniversary of the renewable energy law: legal forces in the fast development of renewable energy' [*kezaisheng nengyuanfa shishi shizhounian:tufeimengjin zhong de fazhi Liliang*], accessed 2 February 2017 at http://www.cnenergy.org/yw/zc/201602/t20160204_269283.html.

Ma, X. and L. Ortolano (2000), *Environmental Regulation in China: Institutions, Enforcement, and Compliance*, Oxford: Rowman & Littlefield Publishers.

National Development and Reform Commission (NDRC) (2011), 'The NDRC Circular on improvement of solar PV on-grid tariff policy', accessed 6 January 2017 at http://www.gov.cn/zwgk/2011-08/01/content_1917358.htm.

Niu, Z. and X. Zhang (2008), 'Reflections on perfecting the renewable energy resource statutes' [*woguo kezaisheng nengyuanfa ruogan wenti wanshan gouxiang*], *Journal of Shandong University of Science and Technology*, 10 (4), 23–7.

PEW Charitable Trusts (2010), report, *Who's Winning the Clean Energy Race?*

Growth, Competition and Opportunity in the World's Largest Economies, accessed 2 March 2017 at www.pewtrusts.org/uploadedFiles/wwwpewtrustsorg/Reports/Global_warming/G-20%20Report.pdf?n=5939.
Porter, M. E. (1998), *The Comparative Advantage of Nations*, with a new introduction, New York: The Free Press.
PRC Amended Law on Renewable Energy (2009), accessed 5 January 2018 at www.npc.gov.cn/npc/xinwen/2009-12/26/content_1538199.htm.
PRC Renewable Energy Law (2005), accessed 5 January 2018 at http://www.npc.gov.cn/englishnpc/Law/2007-12/13/content_1384096.htm.
Renewable Energy Policy Network for the 21st Century (REN21) (2009), *Recommendations for Improving the Effectiveness of Renewable Energy Policies in China*, accessed 6 January 2015 at www.ren21.net/pdf/Recommendations_for_RE_Policies_in_China.pdf.
Renewable Energy Policy Network for the 21st Century (REN21) (2013), *Renewables 2013 – Global Status Report*, accessed 7 July 2016 at http://www.ren21.net/Portals/0/documents/Resources/GSR/2013/GSR2013_lowres.pdf.
Wong, J. (2004), 'The adaptive developmental state in East Asia', *Journal of East Asian Studies*, 4, 345–62.
Wooldridge, A. (2012), ' The visible hand: special report of state capitalism', *The Economist,* 402 (8768).
Yang, M. and R. Pan (2010), *Harvesting Sunlight: Solar Photovoltaic Industry in China*, Background Brief No. 562, Singapore: East Asian Institute.
Yu, Y. (2015), 'Comparison of stimulation policies in wind power and China's application' [*fengdian chanye jili zhengce bijiao yiji zhongguo yingyong tantao*], *Inquiry into Economic Issues* [*jingji wenti tansuo*], 2, 120–25.
Zhang, S., P. Andrews-Speed, X. Zhao, and Y. He (2013), 'Interactions between renewable energy policy and renewable energy industrial policy: a critical analysis of China's policy approach to renewable energies', *Energy Policy*, 62, 342–53.

4. Challenges from the local: geographic and industrial barriers

Despite China's constant effort to expand and diversify energy supply, its coal-dominant energy structure has put it in an awkward position in the context of global climate change. This is furthered by its growing dependence on imported petroleum, which is yet to be fundamentally changed. As part of its national effort for energy self-dependence and against climate change, China's low-carbon electricity strategy is undergoing transformation, which is sensitive to the dynamics of energy policy prioritization, domestic supportive policies, geographic characteristics, and industrial barriers. Variables such as inherent geographic and meteorological conditions, domestic supporting industries, and technological barriers related to power storage and transmission, have constrained the country's renewable energy development.

Aside from the dynamics of domestic supporting industries, China's unique geographic and meteorological features are significant determinants that hamper the realization of policy priorities related to different types of low-carbon electricity generation. China today is facing the world's highest rejection rates for its existing wind and solar power capacity, as a substantial part of the potential energy generated by wind turbines and solar panels in the resourceful west and mid-west areas could not be transmitted to the power grids connecting the economically vibrant coastal areas. Amidst excessive investment powered by massive government subsidies, renewable power generators have been facing the worst curtailment rates in the world, with the national average curtailment ratio in 2016 at 17 percent for wind and 10 percent for solar (Zhou and Lu 2017, p. 2).

GEOGRAPHIC AND METEOROLOGICAL FACTORS

Wind and solar power generation is largely refrained by local meteorological factors such as the intensity of wind and sunshine. As such, most low-carbon electricity projects are sited to adapt to China's land topography, and it is assessed that north and northwest China are suitable for wind and solar. China's prioritization of certain low-carbon electricity technologies is by and large related to the country's inherent geographic and meteorological advantages in tapping energy resources. The fact that China's bountiful coal and hydroelectric resources are mostly located in the northern and western hinterland, and that its wind and solar resources are also concentrated in the vast but sparsely populated northern and northwestern areas, exacerbates the situation of imbalanced distribution of energy resources, as the economically vibrant eastern (coastal) area has high energy demand but is constantly afflicted with energy shortages.

China has abundant onshore and offshore wind resources. According to the fourth national wind energy survey carried out by the China Meteorological Administration in 2014, potential wind energy resources on land can generate an estimated 2,000 GW of wind power at 50 m, 2,6000 GW at 70 m, or 3,400 GW at 100 m above ground given the intensity of wind energy above 300 W/m^2 (Yang et al. 2017, p. 1454).

Some southeastern provinces along the coastlines, including Fujian, Guangdong, Hainan, and Guangxi, also have enormous wind energy potentials, where wind energy intensity can go as high as 600 W/m^2, but such strong winds are difficult to harness as they are often the outcomes of frequent typhoon and tropical depression activities in summer. The most important area in China that is rich in wind resources covers the 'Three North' region, which includes the long inland belt that extends from Xinjiang, Gansu, Ningxia, Qinghai, and Shaanxi provinces in the northwestern part of China, to Inner Mongolia, Shanxi, and Hebei provinces in the north, and then to Liaoning, Jilin, and Heilongjiang provinces in the northeast. Inner Mongolia ranks first in the development of wind power in China, with a technically exploitable capacity of 381.7 GW, followed by Gansu and Xinjiang with 82.2 GW and 64.8 GW, respectively (Yang et al. 2017, p. 1454). In 2016, four provinces, namely Heilongjiang, Jilin, Tibet, and Inner Mongolia, witnessed average wind speeds exceeding 6.0 m/s, with that in Inner Mongolia reaching an astonishing level of 6.63 m/s (China Meteorological Administration 2017, p. 4).

Based on China's climatic and geographic characteristics, Inner Mongolia, Gansu, Xinjiang, Tibet, Jilin, Heilongjiang, and Hebei are primary provinces for wind power development, with Inner Mongolia alone accounting for about one-third of China's total installed wind power capacity. In 2015, the cumulative installed wind capacity in northern China accounted for 32 percent of the nation's total, followed by the northwestern region (30 percent), eastern China (13 percent) and northeast China (12 percent) (Figure 4.1).

As with wind energy, China's vast territorial size endows it with considerable solar PV resources: approximately 60 percent of the land mass enjoys more than 2,200 hours of sunlight a year with solar irradiance

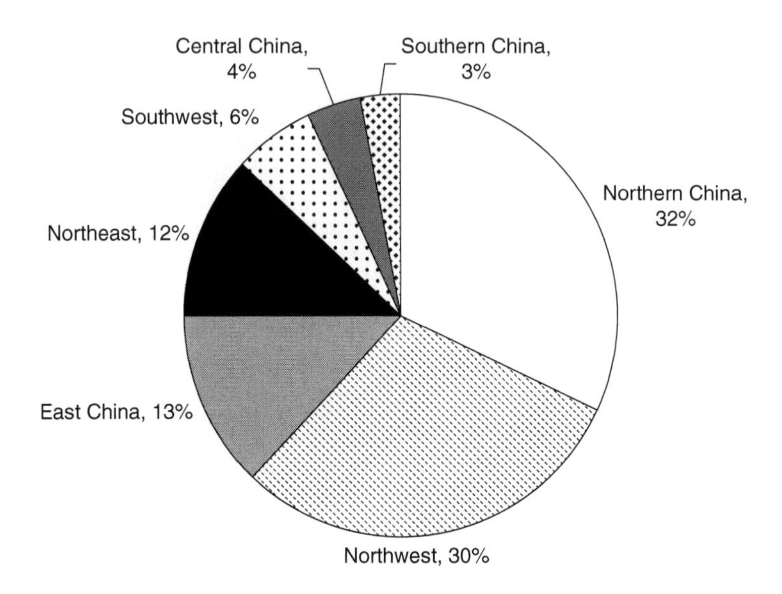

Source: Yang et al. (2017, p. 5).

Figure 4.1 Regional distribution of cumulative wind power installed capacity in China in 2015 (total: 128.53 GW)

greater than 5,000 MJ/m^2 (1,389 kWh/m^2), although some areas may see moderated energy levels due to shielding effects of smog pollution (Deutsche Bank Group 2012, p. 14). Like imbalanced distribution of wind resources, China faces similar geographic constraints in developing solar energy resources. The country's northwestern provinces of Tibet, Xinjiang, and Qinghai and Gansu have the strongest solar radiation that is conducive to solar power generation, but the southeastern provinces, where China's economic powerhouse is, have a much lower solar radiation amount per square meter (Table 4.1).

In 2016, the yearly solar irradiance in northeast, north, northwest, and southwest China exceeded 1,400 kWh/m^2, with eastern Xinjiang Autonomous Region, mid-western Tibet Autonomous Region, most parts of Qinghai province, western Gansu province, and western Inner Mongolia Autonomous Region receiving above 1,750 kWh/m^2 and being the most solar resourceful regions in China (China Meteorological Administration 2017, p. 13). The yearly solar irradiance in some southwestern Chinese provinces like Sichuan and Yunnan as well as in the tropical island of Hainan ranges between 1,400 kWh/m^2 and 1,750 kWh/m^2 (China Meteorological Administration 2017, p. 13). All these provincial regions

Table 4.1 Solar energy resource distribution zones in China

Zone	Annual solar radiation (kWh/m2)	Shares of the national total (%)	Area
I (most abundant)	≥ 1,750	17.4	Tibet, south Xinjiang, Qinghai, Gansu, and west Inner Mongolia
II (very abundant)	1,400~1,750	42.7	Northern Xinjiang, northeast China, east Inner Mongolia, northern China, north Jiangsu, Huangtu Plateau, east Qinghai and Gansu, west Sichuan, Hengduan Mountain, Fujian, South Guangdong, and Hainan
III (abundant)	1,050~1,400	36.3	Hill areas in southeast China, Hanshui river basin, western Guangxi Autonomous Region
IV (normal)	<1,050	3.6	Sichuan and Guizhou province

Source: Li and Wang (2007, p. 24).

are far from China's major industrial bases located in the Pearl River Delta or the Yangtze River Delta.

China's conventional power market has been encountering serious overcapacity and grid transmission bottlenecks for years. China has the largest power generation fleet in the world, with 1,614 GW of installed capacity by 2016, of which over two-thirds was built in the past decade. Despite its gradually slowing down demand, from 2005 to 2015, power capacity grew by an average of 11.5 percent per year. As of the end of 2016, the national power oversupply was 35 percent (Zhou and Lu 2017, p. 9), with those provinces located in western, north, southwest, and northwest China experiencing the most challenging situations. In 2016, only five coastal provinces and cities, Beijing, Hebei, Jiangsu, Shanghai, and Guangdong, exhibited deficits in electricity supply (Table 4.2).

In the context of exacerbating power oversupply, the overconcentration of emerging wind and solar resources in the western part of China, coupled with similar geographic distribution of dominant coal-fired power and hydroelectricity, has imposed enormous pressure on the country's transmission networks, spurring China to launch the immense West–East Electricity Transmission Project, which aims to transmit power from

Table 4.2 Available capacity versus peak demand by province in 2016 (GW)

Province	Available power capacity	Peak demand	Surplus in electricity supply
Anhui	47	33	14
Beijing	10	21	−11
Chongqing	16	16	0
Fujian	40	31	9
Gansu	26	13	13
Guangdong	88	100	−12
Guangxi	30	17	13
Guizhou	38	13	25
Hainan	6	4	2
Hebei	45	53	-8
Heilongjiang	20	11	9
Henan	61	52	9
Hubei	44	29	15
Hunan	31	23	8
Inner Mongolia	75	20	55
Jiangsu	83	92	−9
Jiangxi	22	17	5
Jilin	19	8	11
Liaoning	34	24	10
Ningxia	22	12	10
Qinghai	12	8	4
Shaanxi	30	19	11
Shandong	88	68	20
Shanghai	21	31	−10
Shanxi	61	26	35
Sichuan	51	23	28
Xinjiang	47	26	21
Yunnan	45	14	31
Zhejiang	68	61	7

Source: Zhou and Lu (2017, p. 9).

electricity-rich western regions to the economically prosperous eastern areas. Nevertheless, inefficiency in the northwestern provinces is still astronomically high and the National Development and Reform Commission (NDRC) has realized the urgency of building more ultra-high-voltage transmission lines to carry electricity long distances, and to position new turbines close to major metropolises (*The New York Times* 2017, p. A8).

With the intense competition among thermal power and low-carbon alternatives, including hydropower, wind, nuclear, and solar power, grid companies are in advantageous bargaining positions, which can help them to benefit enormously in such a bottleneck situation. Since most provinces are facing power oversupply, local governments also lobby grid companies to prioritize the transmission of electricity generated by indigenous power plants. Most hinterland provinces are unwilling to voluntarily buy electricity from other provinces, as they themselves are challenged by power overcapacity locally. About 60 percent of the total capacity of the Jiuquan Wind Power Base, one of the world's largest wind farms capable of generating enough electricity to power a small country, goes unused each year (*The New York Times* 2017, p. A8). Part of the problem is the location. Gansu is a barren, mountainous province, chosen for its strong winds and potential for growth. But it is far from the booming cities in eastern China, making transmission of electricity difficult.

Benefiting from such imbalanced distribution, nuclear power plants and offshore wind power plants, which can be built close to coastal centers of demand, and thus save long-distance power transmission costs, have special topographical leverages over hinterland wind and solar power. Nuclear power – despite its controversial impact on the environment and its electricity generating capacity at 15 percent of the world's total – is believed to provide a good path towards a low-carbon electricity future (Rothwell and Graber 2009, p. 176) although previous radiation accidents, such as the Three Mile Island accident in 1979 and the Chernobyl disaster in 1986, raised and caused safety concerns.

Nuclear power plants can be built close to centers of demand, thus saving long-distance power transmission costs. In comparison with other forms of renewable energy, such as wind power and solar PV, nuclear power generation operates at near full capacity and incurs lower costs similar to, or even lower than that of traditional coal-fired power generation. In the case of China, nuclear power is especially important in economically vibrant coastal areas remote from coalfields and sunny and windy places in the western hinterland. Nuclear power accounted for only 1 percent of China's total installed electricity generation capacity and 2.1 per cent of total electricity generated in 2013, which was significantly lower than most industrialized nations (e.g. France 73.3 percent, Britain 18.3 percent, the United States 19.4 percent) and some emerging markets (Brazil 2.8 percent, Argentina 4.4 percent, India 3.5 percent) (International Atomic Energy Agency 2013). China, which used to lag behind in nuclear power development, has been formulating the world's most ambitious energy policies with enormous investment injected to push rapid construction of new nuclear plants.

To alleviate power shortage mainly in the coastal areas, the Chinese government decided to speed up the construction of nuclear reactors to achieve targeted installed capacity of at least 86 GW by 2020, 200 GW by 2030 and 400 GW by 2050. China today has the world's largest nuclear power capacity under construction, both planned and proposed (World Nuclear Association 2011). It is likely to surpass the United States and have the world's largest nuclear power capacity after 2030, and will probably remain in the number one position in this century.

Due to the inconvenient locations of most hinterland wind and solar power plants, China has been making a big bet on offshore wind power projects that are at the doorstep of its coastal industrial bases. The 12th five-year special plan of wind power technology development formulated by the Ministry of Science and Technology included the key technologies research and development of high power wind turbines, such as '10 MW wind turbine overall design technology', '3–5 MW permanent magnet direct drive (PMDD) wind turbine industrialization technology', and '7 MW-class wind turbine development and industrialization technology' (Wu et al. 2014, pp. 454–5). China already had the ability to design and manufacture largescale offshore wind turbines, with hoisting and trial operations for 6 MW offshore wind turbines having been completed.

While the newly installed capacity of China's onshore wind power sector decreased 24 percent in 2016, that of offshore wind power plants saw an increase of more than 60 percent (Liu 2017). China plans to focus on rolling out offshore projects in Jiangsu, Zhejiang, Fujian, and Guangdong provinces, with the capacity of the facilities already under construction in the four provinces expected to surpass 1 GW by 2020. The total installed capacity of facilities under construction across the entire country is expected to reach 10 GW by 2020, with total on-grid capacity hitting 5 GW (Liu 2017). Offshore wind power plants have the potential to become formidable competitors against onshore renewable projects because not only can they utilize the abundant natural resources, but they can also relieve the power shortages in coastal China (Chen 2011, p. 5013). Nevertheless, compared with onshore wind power, the development of offshore wind power is still facing difficulties in building power transmission, in harnessing the harsh offshore natural environments, and fulfilling multisectoral coordination and management (Wu et al. 2014, p. 453).

INDUSTRIAL BARRIERS: GRID RESTRICTIONS AND UNUSED CAPACITY

Wind and solar power, though inexhaustible in quantity, is highly variable and difficult to predict due to the intermittent and fluctuating nature of these renewable energy resources. Wind and solar only generate electricity when the wind is blowing or the sun is shining, and the energy produced from these renewables fluctuates on a daily basis according to the varying meteorological conditions. The wind power generation system exhibits a significant and drastic variation in output power, which is fundamentally not connected to the demand for electrical power. Intermittent wind and solar cannot be stockpiled in the absence of energy storage systems and must also be used when available, as otherwise they lose energy potentials (Evans et al. 2012, p. 4141).

The abundant renewable energy may have to be curtailed because real-time balance between load and generation must be maintained, and electric generation cannot be economically stored on a large scale. In addition, the renewable energy generated cannot be utilized for equipment maintenance, upgrade works or failure. (Li et al. 2015, p. 1068). As a result of wind and solar developers' rush-to-build actions driven by the government's supportive policies, the surge in China's installed renewable energy capacity has led to huge inefficiencies, with the world's highest renewables curtailment rates, standing at 13.6 percent and 15.5 percent for wind and solar, respectively, in the first half of 2017 (Zhou and Lu 2017, p. 23). In addition to long transmission distances between many renewable energy projects and the demand side, other industrial and institutional barriers including the ineptness of old grid systems at incorporating intermittent wind- and solar-generated electricity (Wang 2010, p. 708), underdeveloped energy storage technology, and low contract implementation rates have hampered the installed capacity from being utilized efficiently.

The power grid networks in many countries are aging, designed around the concept of large, controllable electric generators. Usually the grid operator uses a three-phase planning process to ensure power plants produce the right amount of electricity at the right time to consistently and reliably meet electric demand. Because the grid has very little storage capacity, the balance between electricity supply and demand must be maintained at all times to avoid a blackout or other cascading problem (Fares 2015). Intermittent renewables are challenging because they disrupt conventional methods for planning the daily operation of the electric grid. Their power fluctuates over multiple time horizons, forcing the grid operator to adjust its day-ahead, hour-ahead, and real-time operating procedures (Fares 2015). Today, the intermittency problem associated with the phasing in of low-carbon sources of energy such as solar and wind power is not unique to China. Even Germany, with one of Europe's most stable grids, has difficulty maintaining grid stability with renewables supplying 22 percent of electricity demand (Ruud 2014, p. 176).

In China, however, the inability of the grid system to incorporate intermittent renewable electricity has become a more severe bottleneck due to the lack of competitiveness among state-owned grid companies, overinvestment in renewable power projects, the overcapacity of China's power industry as a whole, and remote locations of many wind and solar power plants. Power generated from renewable sources decreases accuracy of load forecast and therefore affects power grid dispatching and operation, impacting frequency control of the power grid, voltage regulation, power supply quality, and fault levels and stability of the power grid (Wang and Chen 2004, pp. 49–53; Wang 2010, p. 708). Moreover, the sluggish construction and development of China's power grid cannot keep up with the surging installed capacity of renewable energy, especially in the sparsely populated western and northern areas. From 2002 to 2007, investment in generation capacity increased from 74.7 billion yuan (US$9.0 billion) to 322.6 billion yuan (US$42.2 billion) at an average annual rate of 28 percent, but investment in the grid system increased at a year-on-year rate of only 9 percent, from 157.8 billion yuan (US$19 billion) to 245.1 billion yuan (US$32.3 billion) (Wang 2010, p. 709). Despite the mandatory grid connection for renewable energy under the Renewable Energy Law (2005) and its amendment (2009), grid operators, which are state-owned monopolies in China, still have little motivation to expand service to match the fast development of renewable energy.

As China has a fragmented seven-grid system, comprising the Northeast China Grid, North China Grid, Northwest China Grid, East China Grid, South China Grid, Tibet Grid, and Central China Grid, the inter-grid electricity exchange capacity has been vital for transmission and distribution of renewable energy over long distances. Nevertheless, the aging grid system has resulted in insufficient inter-grid exchange capacity and underinvestment in smart grid, an ideal solution to the intermittency problem associated with renewables. China's average wind curtailment rate stood at a record high of 15 percent in 2015, with rates soaring to over 30 percent in wind-rich provinces such as Gansu, Xinjiang, and Jilin (Table 4.3). Up to 33.9 billion KWh of wind electricity failed to connect to the grid, leading to losses of US$2.8 billion (Li 2016). The northwestern provinces of Gansu and Xinjiang saw some of the worst grid congestion for both wind and solar. In 2016, Gansu's wind and solar curtailment surged to 43 percent and 30 percent respectively, while Xinjiang's curtailment rate remained above 30 percent. China's 2016

Table 4.3 Wind power curtailment rates in wind-rich provinces in 2015

Province	Curtailment rates of wind power	Wind capacity of % of all installed capacity integrated into the grid in 2015
Gansu	39%	26.97%
Xinjiang	32%	26.07%
Jilin	32%	17%
Heilongjiang	21%	19%
Inner Mongolia	18%	23.31%
Ningxia	13%	26.03%
Hebei	10%	17.69%
Liaoning	10%	14.78%
Shanxi	2%	9.6%
Qinghai	0%	2.27%
Tibet	0%	0.51%

Source: Li (2016).

national average curtailment ratios for wind and solar amounted to 17 percent and 19.8 percent respectively (Zhou and Lu 2017, p.22). As curtailment goes uncompensated, China's renewable energy asset owners face huge financial risks.

China has decided to speed up electricity grid construction to solve the high curtailment issue. The Chinese government once put its strategic focus on ultra-high-voltage (UHV) power transmission rather than smart grids. To transfer electricity generated in northwestern provinces to energy-hungry coastal areas in the east, long distance transmission capacity is needed. China has decided to upgrade transmission lines and build the world's most capable UHV transmission network that can transmit 1,000 kV or more over long distances. A total of 12 such UHV transmission lines are being built to connect the northwestern provinces and coastal areas. The world's longest UHV line will run from the Changji Prefecture in Xinjiang to Xuancheng city in east China's Anhui province. The 3,324 km line is designed to accommodate currents of 1,100 kV. However, it usually takes at least three years to build a UHV line, which is also very costly. Intra-province integration within local grids is a more feasible solution.

Local distribution grids can be converted into smart grids by implementing the latest technology, which helps integrate more renewable energy while avoiding costly transmission investments. Unfortunately, due to its preference of UHV lines over smart grids, China's investment in smart grid infrastructure had been sluggish until recently. It has been estimated that China will invest US$77.6 billion in smart grid infrastructure between 2016 and 2026 (PRNewswire 2016).

OVERCAPACITY IN AN ERA OF EXCESSIVE INVESTMENT

In its 'renewable energy revolution' (Mathews and Tan 2015) driven by government support and stimulation, China has become the world's largest renewable energy developer in terms of cumulative installed capacity of wind and solar photovoltaics (PV) (Dent 2015; Chen 2016). Behind such glamorous figures, however, are numerous idled wind turbines and solar panels installed on the country's vast hinterland. Chen (2016, p. 14) looked at China's renewable energy progress through the lenses of the development state (Amsden 1992; Weiss 2003) and ecological modernization, two competing modes of governance that Chinese policy elites seek to reconcile. The author remained optimistic about China's state capacity of paying equal attention towards the ecological and economic aspects of modernization, arguing that pervasive development-centric ideology in a developmental state model can be restrained by an ecology-focused central government, through mechanisms deployed to minimize the constant problems of shirking and slippage at the local level. In other words, using the framework of principal–agent theory, the center-as-principal is able to manipulate incentives and supervise locals-as-agents in the process of recentralization aimed at achieving sustainable development. Yet, in reality, overcoming such a principal–agent dilemma to strike a balance between quantity and quality, supply and demand, economy and environment has never been an easy job, a point manifested by China's extremely high renewables curtailment and the huge financial risks of such investment.

The severity of excessive investment and overcapacity in renewable energy has put pressure on China's National Energy Administration (NEA) to launch an investment risk alert mechanism for both solar PV and wind power generation. Updated annually, the mechanism assigns one of three ratings – red (high risk)/orange (medium risk)/green (low risk) to each region. The NEA's wind power assessment in 2018 flagged Gansu, Xinjiang, and Jilin provinces with red alerts, where all the wind power generation projects, including those that had been approved by the authorities, had to be suspended (National Energy Administration of China 2018). Inner Mongolia, Heilongjiang, and parts of Shanxi, Hebei, and Shaanxi provinces were labeled as orange alerts areas, where no new wind power projects would be approved by the end of the 2018. The NEA also released a set of investment risk evaluation metrics for utility-scale PV in August 2017. Only eight of the 38 regions evaluated were considered low risk (green); 24 were medium-risk (orange), and six were high-risk (red) (Zhou and Lu 2017, p. 19).

Geographic mismatches, intermittency of renewable energy and grid congestion are only typical industrial barriers that have amplified China's across-the-board problem of industrial inefficiency and overcapacity in the emerging renewable sector. In China's model of state capitalism, a partially marketized developmental structure evolving from the previous planning of the economy and obsession with capacity overexpansion powered by excessive investment has been permeating all layers of the state and the market. Motivated by all kinds of quantitative growth targets from both the central and local governments, market players and other stakeholders always prioritize quantity more than quality, and pomp rather than efficiency. The central government's renewable energy targets for 2015–30 included the pursuit of 11.4 percent of non-fossil fuels in total energy mix by 2015, 15 percent by 2020, and 20 percent by 2030, and wind and solar power capacity of 100 GW and 21 GW by 2015, 150 and 70 GW by 2017 and 200 and 100 GW by 2020, respectively (Mathews and Tan 2015, pp. 32–3). As most of these quantitative targets emphasize installed capacity rather than the efficiency issue, high curtailment and huge waste became inevitable in the great leap forward in renewable energy.

Incentivized by grants, subsidies, and tax rebates, Chinese energy developers often focus on the number of wind turbines or solar panels installed in a specific project, ignoring the accessibility of power grids or the long distances between supply and demand. Sometimes, the government's renewable energy strategy also has a political or social function. With a political campaign of poverty alleviation on track, the Chinese government often intentionally promotes renewable energy projects in some remote and poverty stricken areas, where social function is remarkable while economic efficiency is rather low (Zhang et al. 2009, p. 446).

Grid access is crucial for the commercial success of renewables, but conflicts between renewable power generators and grid companies have been particularly serious in China, especially with the growing scale of renewable energy capacity (Wang 2014, pp. 214–15). Both at the central and local levels, government departments and grid dispatchers with vested coal interests have been accused of exerting administrative power to disrupt wind and solar energy production. Like many other countries, China's response to energy challenges is intertwined with its energy policy process, in which even the powerful NDRC lacks the authority, resources, and tools to prevail over other state energy oligarchies. State-owned enterprises (SOEs) that dominate China's fossil fuel supply as well as thermal power generation and transmission sectors are major forces blocking local renewable energy production.

In the coal production industry, large-sized SOEs, most of which are supervised by provincial governments, account for about half of China's domestic coal output. The remaining half of the country's coal output comes from small-scale SOEs supervised by local governments and numerous privately or collectively owned coal mines in villages and townships. China's coal production industry has been partially liberalized to allow participation of privately invested mines, thus creating a large number of multi-millionaires, whose interests are often threatened by the booming local renewable energy projects. Coincidentally, many coal-rich provinces like Inner Mongolia, Shaanxi, and Xinjiang happen to be regions with abundant solar and wind resources. With competition for grid connection becoming fierce among different types of energy, grid companies that have traditional ties with coal-fired thermal plants often play vital roles in determining winners and losers.

When designing renewable energy policies, Chinese policy-makers expected local governments to play a very cooperative role in enforcing those policies, laws, and regulations. In practice, however, this high degree of administrative cohesion does not exist in places where local government interests diverge sharply from those of the central planners and create substantial obstacles to strict enforcement of both national and local energy policies. As stakeholders of local fossil fuel companies, local authorities in many cases tend to help these companies bypass strict energy directives and curb competition from local renewable players.

China's weakness in enforcing national energy policies is subject to the level of coordination within three kinds of relationship, namely the relationships between central government agencies, the relationships between central and local governments and the relationships between SOEs and government agencies. First, responsibilities are not clearly defined among central government agencies and long bargaining process emerges when conflict of interests occurs. Second, the multi-layer local regimes have great scope for distortion or non-implementation of national energy policy (Andrews-Speed 2003, p. 53). Third, interest groups such as the gigantic energy SOEs in China's energy sectors have gained more influence in policy implementation. The overconcentration of both renewable and conventional power resources in some provinces has exacerbated the tension and competition among them, leading to even less accessibility of wind and solar power to grid.

REFERENCES

Amsden, A. H. (1992), *Asia's Next Giant: South Korea and Late Industrialization*, Oxford: Oxford University Press.

Andrews-Speed, P. (2003), *Energy Policy and Regulation in the People's Republic of China*, The Hague: Kluwer Law International.

Chen, G. C.-F. (2016), *Governing Sustainable Energies in China*, New York: Palgrave Macmillan.

Chen, J. (2011), 'Development of offshore wind power in China', *Renewable and Sustainable Energy Reviews*, 15 (9), 5013–20.

China Meteorological Administration (CMA) (2017), *China Wind and Solar Energy Resources Bulletin 2016* [*zhongguo fengneng taiyangneng ziyuan nianjing gongbao 2016*], Beijing: CMA Wind and Solar Energy Resources Center.

Dent, C. M. (2015), 'China's renewable energy development: Policy, industry and business perspectives', *Asia Pacific Business Review*, 21 (1), 26–43.

Deutsche Bank Group (2012), *Scaling Wind and Solar Power in China: Building the Grid to Meet Targets*, Deutsche Bank Group report, accessed 1 March 2018 at file:///C:/Users/eaicg/Downloads/Scaling%20Wind%20and%20Solar%20Power%20in%20China_Building%20the%20Grid%20to%20Meet%20Targets.pdf.

Evans, A., V. Strezov, and T. J. Evans (2012), 'Assessment of utility energy storage options for increased renewable energy penetration', *Renewable and Sustainable Energy Reviews*, 16, 4141–7.

Fares, R. (2015), 'Renewable energy intermittency explained: challenges, solutions, and opportunities', *Scientific American*, accessed 1 March 2017 at https://blogs.scientificamerican.com/plugged-in/renewable-energy-intermittency-explained-challenges-solutions-and-opportunities/.

International Atomic Energy Agency (IAEA) (2013), 'Nuclear share of electricity generation in 2013', accessed 4 October 2014 at http://www.iaea.org/PRIS/WorldStatistics/NuclearShareofElectricityGeneration.aspx.

Li, C., H. Shi, Y. Cao, J. Wang, Y. Kuang, Y. Tan, and J. Wei (2015), 'Comprehensive review of renewable energy curtailment and avoidance: a specific example in China', *Renewable and Sustainable Energy Reviews*, 41, 1067–79.

Li, J. and S. Wang (2007), *China Solar PV Report*, Beijing: China Environmental Science Press.

Li, Y. (2016), 'Blowing in the wind', *China Dialogue*, 31 May 2016, accessed 1 March 2017 at https://www.chinadialogue.net/article/show/single/en/8965-Blowing-in-the-wind.

Liu, Y. (2017), 'China makes a big bet on offshore wind', *Renewable Energy World*, accessed 2 March 2018 at http://www.renewableenergyworld.com/articles/2017/10/china-makes-a-big-bet-on-offshore-wind.html.

Mathews, J. A. and H. Tan (2015), *China's Renewable Energy Revolution*, New York: Palgrave.

National Energy Administration of China (NEA) (2018), 'NEA circular on monitoring and warning of wind power investment in 2018' [*guojia nengyuanju guanyu fabu 2018niandu fengdian touzi jiance yujing jieguo de tongzhi*], NEA document, 23, 5 March 2018.

PRNewswire (2016), 'China smart grid: market forecast (2016–2026)', accessed

1 March 2018 at https://www.prnewswire.com/news-releases/china-smart-grid-market-forecast-2016-2026-300368356.html.
Rothwell, G. and R. Graber (2009), 'The role of nuclear power in climate change mitigation', in F. P. Sioshansi (ed.) *Generating Electricity in a Carbon-Constrained World*, Oxford: Academic Press, pp. 176–9.
Ruud, A. (2014), 'Smart, but is it sustainable? The importance of reconciling non-technical concerns in grid-development policies', in E. Moe and P. Midford (eds), *The Political Economy of Renewable Energy and Energy Security: Common Challenges and National Responses in Japan, China and Northern Europe*, New York: Palgrave Macmillan, pp. 175–93.
The New York Times (2017), 'Windmills stand idle in China as even more are being constructed', 16 January 2017, p. A8.
Wang, Q. (2010), 'Effective policies for renewable energy – the example of China's wind power – lessons for China's photovoltaic power', *Renewable and Sustainable Energy Reviews*, 14 (2), 702–12.
Wang, W. and M. Chen (2004), 'Towards the integrating wind power into power grid in China', *Electricity*, 53, 49–53.
Wang, Y. (2014), 'A review of renewable energy legislation and policies in China', in E. Moe and P. Midford (eds), *The Political Economy of Renewable Energy and Energy Security: Common Challenges and National Responses in Japan, China and Northern Europe*, New York: Palgrave Macmillan, pp. 197–220.
Weiss, L. (2003), 'Guiding globalisation in East Asia: New roles for old developmental states', in L. Weiss (ed.), *States in the Global Economy: Bringing Domestic Institutions Back In*, Cambridge: Cambridge University Press, pp. 245–70.
World Nuclear Association (2011), 'World nuclear power reactors and uranium requirements (2011)', accessed 8 January 2016 at http://www.world-nuclear.org/info/reactors.html.
Wu, J., Z. Wang, and G. Wang (2014), 'The key technologies and development of offshore wind farm in China', *Renewable and Sustainable Energy Reviews*, 34, 453–62.
Yang, J., Q. Liu, X. Li, and X. Cui (2017), 'Overview of wind power in China: Status and future', *Sustainability*, 9 (8), 1454.
Zhang, P., Y. Yang, J. Shi, Y. Zheng, L. Wang, and X. Li (2009), 'Opportunities and challenges for renewable energy policy in China', *Renewable and Sustainable Energy Reviews*, 13 (2), 439–49.
Zhou, Y. and S. Lu (2017), *China's Renewables Curtailment and Coal Assets Risk Map*, Beijing: Bloomberg New Energy Finance.

5. Interest groups and the bureaucracy

The crux of the energy policy dilemma, in the view of many political observers, is an emergent conflict between a well-established set of segregated policy subsystems and a diversity of new demands for a policy that requires considerable interest integration (Chubb 1983, p. 4). As reliable and sufficient energy supply at an affordable price is strategically important for any country's national economy, and most energy projects involve high capital investment and large risks, almost every government in the world realizes it necessary to formulate a comprehensive system of energy policies that cover the production, transformation, transportation, and consumption of energy.

Like many other countries, China's energy policy-making is fueled by the country's overall economic growth targets. Given the profound impact on a broad set of interest groups in the party-state, the energy policies are often the outcomes of bureaucratic politics and inter-agency competitions among various interest groups rather than reasoned debates.

CHINA'S FRAGMENTED ENERGY BUREAUCRACY

China's renewable energy policy is part of its energy planning, which has long been deemed by the government as a pillar of China's overall economic development strategy. Given this outlook, renewable energy policy-making has always fallen into the area over which the National Development and Reform Commission (NDRC), the country's economic planning ministry, has the most authority. Similar development-centric ideology had allowed the NDRC to dominate other climate change-related policies until March 2018, when its long-time bureaucratic rival, the Ministry of Environmental Protection, was enhanced and restructured by the top leadership into the Ministry of Ecological Environment, which then took the power away from the NDRC on issues such as climate change and greenhouse gas emissions (*The Straits Times* 2018).

Inter-agency conflicts between the NDRC and Ministry of Environmental Protection became obvious as domestic environmental problems had exacerbated over the past decade (Chen 2009, p. 22), reflecting ideological disputes within the Chinese bureaucracy over the balance between economic growth and environmental protection. Despite the 2018 governmental reshuffle in favor of pro-environment apparatus, renewable energy development is still being scrutinized by the central government mostly as an energy issue rather than a mitigation tool from the climate change perspective. The new Ministry of Ecological Environment has taken charge of formulating and implementing policies, plans, and standards related to carbon emissions cuts and other pollution activities of energy enterprises, but it still lacks authority over renewable energy development, which has been entrenched in the energy administration's policy-making.

Issues in Chinese politics are usually organized vertically in what are called *xitong* (systems) (Bachman 1998). The hierarchical components of the 'energy system' (*nengyuan xitong*) include the National Energy Commission (NEC) at the top, which formulates energy strategy and coordinates relevant ministries, departments, and entities. This is followed by the NDRC and its semi-independent National Energy Administration (NEA). These agencies are in charge of detailed energy policy formulation, daily coordination and secretarial work inside the NEC and long-term strategic planning for national energy development. However, representatives of the NDRC and NEA have often failed to dominate the decision-making process inside the NEC, due to the presence and influence of powerful players like the Ministry of Finance, the Ministry of Transport, the Ministry of Commerce, and state-owned energy giants such as national oil companies (NOCs), the State Grid Corporation and major coal companies (Interview with researchers from the NDRC Energy Research Institute, 2 March 2012, Beijing).

Although basic staff support for the NEC is provided by the NDRC and NEA, the NEC institutional members include almost all the cabinet ministries inside the State Council and other relevant stakeholders. China's energy policy-making process is, to a large extent, vertically organized, with Party leaders in charge of the energy system (*nengyuan xitong*) at the top having the last word on all vital issues. However, as final arbiters of energy policy-making, the influential Party leaders tend to become consultative and consensual due to the complexity of energy issues. Its diversified components have revealed that China's top-level energy policy-making is a complicated process that absorbs viewpoints from various institutional players as opposed to being exclusively managed by energy professionals. In recent years, although non-governmental organizations (NGOs) and the emerging civil society have become visible players, their roles and functions are still marginal under China's state-centric political systems.

The maintenance of high economic growth has been the strategic priority of the Chinese government. Hence, the NDRC that plans the country's economic development has been playing a pivotal role in shaping the country's energy policy. Despite this, energy planners do not, however, have sufficient authority, resources, and tools to prevail over other state energy oligarchies. Evidently, China's energy policy formulation does not fit in the rational decision-making model where parochial interests often give way to overall national interest. The concept of 'fragmented authoritarianism' identified by Kenneth Lieberthal (1992) in China's decision-making provides a useful lens for viewing the processes of formulating and implementing energy-related policies.

An authoritarian centralist state in theory, in the past three decades China has witnessed the emergence of de facto federalism and plutocracy in practice. Since then, its energy policy-making has been undergoing dramatic changes such as pluralization, decentralization, and fragmentation. The surge in China's energy supply and demand stems from the massive industrialization process. While this has in fact further integrated it into the global system, it has led to a subsequent increase in the number and types of pressure groups involved in energy policy-making. This is likely to expand substantially to include more ministries at the national level, big business entities, media, local governments, and NGOs among other entities and individuals. All provinces have their own regulatory departments to deal with energy issues and many large state-owned enterprises (SOEs) have substantial stakes in energy businesses. Recently the influence of the media and think-tanks over the energy policy process has been increasing, while municipal and even county-level authorities have become more active in vying for favorable energy-related policies.

Like many other countries, China's response to energy challenges is intertwined with its energy policy process, in which even the powerful NDRC lacks the authority, resources, and tools to prevail over other state energy oligarchies. Since the past 30 years, China's energy bureaucracy has undergone periodic restructuring in order to establish a central energy agency aimed at ending the splintering of energy sector authority among multiple institutions. It has, however, been unable to do so despite the growing importance of energy security in China's domestic and foreign agendas. The high-level NEC headed by two vice premiers was established in 1980 but it was abolished merely two years later. A Ministry of Energy was established in 1988 but was eliminated again in 1993, giving way to market allocation of energy resources. The low-rank and understaffed Energy Bureau was set up subsequently under the NDRC but was unable to effectively coordinate industries and ministries in the formation and enforcement of energy policies. As opposed to hundreds, if not thousands, of employees working in the energy bureaucracies of many developed nations, the NDRC's Energy Bureau in China had only about 30 employees. In recognition of the need to strengthen energy sector management, the government in 2005 established the National Energy Leading Group under the State Council. This was an inter-agency coordination panel headed by Premier Wen Jiabao himself. During another round of governmental restructuring in 2008, the National Energy Leading Group was formalized into the National Energy Commission, while the NDRC's Energy Bureau was upgraded to the vice-ministerial-level National Energy Administration.

This bureaucratic restructuring was merely about putting old wine in new bottles and it did not fundamentally change the scattered responsibility and authority among multiple powerful stakeholders. Turf wars among various energy institutions have often resulted in the formation of energy laws that fail to specify which agencies have been responsible for the content of those laws, delaying or preventing implementation. Consequently, China's state-owned energy companies are powerful and relatively autonomous due to their full or vice-ministerial ranking. They often have the ability to initiate major energy projects and policies and sometimes advance corporate interests at the expense of national ones (Downs 2008b, p. 42).

Since the high-profile governmental reorganization in 2008, both the NEA and the Ministry of Environmental Protection were entitled to more political clout than their predecessors, but this was still not enough to mediate the bureaucratic infighting that disrupts energy and climate policy-making. The heads of some of the politically powerful and relatively autonomous state-owned energy companies – for example, the China National Petroleum Corporation (CNPC), China Petrochemical Corporation (Sinopec), State Grid Corporation, and the Shenhua Group – all hold ministerial ranks (Downs 2008b, p. 43). As successors to the former industrial ministries (e.g., the Ministry of Petroleum and the Ministry of Electrical Powers) in the period of the command economy, many of these energy giants still see their executives sit in the elite Communist Party of China (CPC) Central Committee with access to face-to-face discussions with top leadership. Conflicts in the objectives of the energy policy within itself and with other polices (e.g., climate-related policy) are thus unavoidable (Andrews-Speed 2004, p. 54). For instance, the state-owned petroleum, coal, hydropower, wind, solar, and nuclear giants all intend to increase their production and obtain more subsidies or tax rebates from the government. Simultaneously, however, the central government has had to reduce reliance on traditional fossil fuels including coal and petroleum and increase the share of clean energy such as hydropower, wind power, and solar power for the sake of sustainable development and mitigation of climate change. Chinese officials and energy experts are well aware of the need to make the country's energy institutions effective and there is growing support for the establishment of a ministry of energy. Despite this, powerful ministerial and corporate interests favor the status quo, with energy firms being reluctant to have another political manager, fearing that it would limit their direct access to China's leadership (Downs 2008b, p. 45).

'TIGERS' IN CHINA'S ENERGY REALM

Gigantic SOEs dominate China's fossil fuel supply as well as power generation and transmission sectors. In the coal production industry, large-sized SOEs, most of which are supervised by provincial governments, account for about half of China's domestic coal output. Yet the Shenhua Group and the China National Coal Group (China Coal), the two largest coal SOEs in the country, are still under the auspices of the central government. The remaining half of country's coal output comes from small-scale SOEs supervised by local governments and numerous privately or collectively owned coal mines in villages and townships.

In contrast to the country's oil exploitation sector that is still the exclusive domain of the three NOCs, China's coal production industry has been partially liberalized to allow the participation of privately invested mines, thus creating a large number of multi-millionaires. Compared with the boom of private coal mines since 1990s, the market share of small and medium-sized SOE mines supervised by local government showed a declining trend in recent years. In 2009, such local SOE mines accounted for only 12 percent of national coal output (Tu 2011, p. 7). The liberalization and privatization process, however, does not prevent state-owned coal giants, especially the Shenhua Group and China Coal, from playing a dominant role in the market as well as in the policy-making process. Like their peers in China's oil industry, both Shenhua and China Coal were encouraged by the Chinese government to set up subsidiaries into which they virtually transferred their core assets and got listed on the domestic and overseas stock exchanges.

China Shenhua was listed on the Hong Kong Stock Exchange in June 2005 and on the Shanghai Stock Exchange in October 2007, given that it managed the production and sales of coal, railway, and port transportation of coal-related materials, as well as power generation and sales. The China Coal Energy Company was listed on the Hong Kong Stock Exchange in December 2006 and on the Shanghai Stock Exchange in February 2008.

With the large amounts of funding raised through initial public offerings (IPOs) and the government's forceful policies to support large-scale and efficiently operated coal mines, since then both China Shenhua and China Coal have witnessed rapid expansion.

China's coal mining industry is the largest and also the most dangerous, with thousands of miners losing their lives to coal pit accidents. Lobbied by Shenhua, China Coal, and other giant SOEs, the Chinese government has been formulating policies based on production safety and environment protection to close down small-sized private coal mines, which often ended in mergers and acquisitions by large SOEs. Coal-burning contributes to more than two-thirds of China's electricity output. These coal mine giants also have huge influence in the power sector. Shenhua, as an example, controlled and operated 55 coal-fired power units, 3 gas-fired power units and 21 wind power units, with a total installed capacity reaching 23,520 MW, up by 25.1 percent on a yearly basis by the end of 2009. In 2009, Shenhua's total generation capacity amounted to 105.09 billion KWh, up 7.5 percent per year (China Shenhua 2012).

In the area of exploitation of crude oil, natural gas, and downstream refining, three state-owned NOCs, that is, the China National Petroleum Corporation (CNPC), China Petrochemical Corporation (Sinopec) and China National Offshore Oil Corporation (CNOOC), are the officially authorized oligopolists that have reaped massive profits from China's surging oil and gas demand since the 1990s. As far as low-carbon energy development is concerned, these NOCs support less-polluted natural gas as an alternative rather than renewables, including wind and solar power. Compared with the hefty roles played by coal and crude oil in China's energy mix, natural gas has been a minor fossil fuel player in China's energy mix for decades. The rapid urbanization and industrialization of the past ten years have resulted in a steady increase the percentage of natural gas in China's total energy production and consumption. The Chinese government perceives natural gas as a cornerstone of its energy policy over the next decade because it burns much cleaner than coal and is relatively plentiful within China's territory. China's natural gas demand will probably grow at 15 percent annually over the next decade and making it a more important market globally. China, the world's third biggest user of natural gas, after the United States and Russia, has to import around 40 percent of its total needs as domestic production cannot keep up with demand. China's 2017 imports of pipeline gas and liquefied natural gas (LNG) topped 67 million tonnes, up by more than a quarter from a year earlier, with LNG imports alone surging more than 50 percent (CNBC News 2018).

Although natural gas-fired power plants have the advantages of high efficiency and low emissions, the percentage of China's total power generation derived from natural gas has been less than 1 percent. About one-fifth of the total natural gas demand is consumed by urban households mainly for heating and cooking, and this ratio is expected to surge as China's urbanization rate continues to increase. Like their peers in other countries, the three Chinese NOCs have a close and interlocking relationship with the Chinese government. They fulfill important strategic and socioeconomic functions such as securing oil supply from both home and abroad for the sustainable development of the national economy and stabilizing the price level of processed oil to appease domestic consumers. China underwent major bureaucratic streamlining in the 1980s to separate government departments from commercial corporations with an intention to increase efficiency.

Subsequently, the oil production, refining, and investment functions pertaining to the Ministry of Petroleum and Ministry of Chemical Industry were gradually transferred to the newly established NOCs. The State Council set up the CNOOC in 1982 to cooperate with foreign investors in the exploitation of offshore oil and gas resources, and the latter was soon given monopoly status in offshore oil and natural gas exploration. A year later, Sinopec was established from the downstream assets (refining and marketing) of the Ministry of Petroleum and the Ministry of Chemical Industry (Downs 2008a, p. 122). In 1988, the central government disbanded the Ministry of Petroleum and established the CNPC to handle all petroleum activities in China. Originating from government departments, the three NOCs hold a rank equivalent to a vice-ministerial one and hence possess major political privilege when dealing with other ministries.

The general managers of the NOCs are directly appointed by the Party's Central Organization Department, some of whom are members in the Party's powerful Central Committee, a political club of officials at ministerial or even higher ranks. Through the revolving door mechanism manipulated by the Party's Central Organization Department, the chief executives in the NOCs may be transferred to serve as governor or vice governor of a province, and the process is reversible. Former CNPC General Manager Jiang Jiemin for example was transferred from the post of CNPC vice president to serve as vice governor of Qinghai province in 2000. After four years, Jiang was moved back to CNPC and served as its deputy general manager. Su Shulin, former president of Sinopec and an alternate member of the 17th CPC Central Committee became Governor of Fujian Province in 2011. Some officials with petroleum industry backgrounds were even given membership of the CPC Politburo Standing Committee, the most powerful sanctum inside the party-state possessing the highest authority. Zeng Qinghong, a CNOOC executive who finally joined the Politburo Standing Committee in 2002, was China's vice president from 2003 to 2008 and regarded as the head of the princeling faction inside the top leadership. Zhou Yongkang, a CNPC veteran and confidante of former president Jiang Zemin, entered the Politburo Standing Committee in 2007 to head political and legal affairs and command the country's armed police force. In the anti-corruption campaign launched by CPC General Secretary Xi Jinping in 2012, Zhou and many of his protégés in the petroleum sector, including Jiang and Su, were jailed under graft charges.

Although the party-state still has a firm grip over the NOCs through majority ownership and personnel appointment, these companies – especially after their subsidiaries were listed on overseas and domestic stock exchanges in the late 1990s – have become more autonomous and profit-driven than before. When their own commercial interest, home and abroad, conflicts with the regime's overall interests, these oil giants tend to use their accumulated political leverage – based on China's surging oil reliance – to their benefit. This is unfortunate in situations where regime interests often take political, social, and diplomatic factors into account. In the process of marketization and globalization, leading SOEs in key sectors that are vital for China's economic security have been upheld by the central government as national champions. Their rapidly expanding economic interests might reinforce this policy support and reconstruct their relationship with the government through powerful lobbying and buy-off activities.

The enormous financial, political, and personnel resources these NOCs can mobilize in cases when their interests do not coincide with those of the party-state have intensified the fragmentation of China's energy policy-making process. This has only exacerbated the power diffusion that has become a salient feature of the post-Deng Xiaoping era. The growing economic strength and political influence of these NOCs have encroached on the central government's authority over overall energy policy-making. The NOCs are sometimes depicted by domestic media as entities with monopolistic vested interests that prioritize profits over social and environmental well-being, and have come to be regarded as an interest group for a number of reasons: their subsidiaries are listed on foreign stock exchanges, their profits are growing, the senior management is becoming internationalized, and their role in securing oil imports and their relative strength vis-à-vis the central government's energy bureaucracy is increasing (Downs 2008a, p. 125).

The main businesses of China's non-fossil fuel sectors are all related to the power industry, which has become increasingly competitive as a result of government-initiated structural reforms aimed at increasing competition and efficiency. In December 2002, the State Council broke the State Power

Corporation's monopoly over the power industry by dismantling the corporation into 11 smaller companies: two electric power grid operators, five electric power generation companies, and four consultant and construction companies. Previously, the ministerial-level State Power Corporation had controlled half of the nation's electricity generating assets and almost all the transmission grids. A new ministry-level body, the China Electricity Regulatory Commission (CERC), was established thereafter to oversee the deregulation and power industry. The Big Five generators, namely China Huaneng, China Datang, China Huadian, Guodian Power, and State Power Investment, obtained an equal share of the generating assets of the State Power Corporation. The Big Five today contribute to about half of China's power output. On the transmission front, the Beijing-based State Grid Corporation of China acquired most of the grids asset from the defunct State Power Corporation, which transferred part of its power grids to the Southern Power Grid Corporation headquartered in Guangzhou. The Southern Power Grid manages the grids in the southern provinces and regions of Guangdong, Hainan, Guangxi, Yunnan, and Guizhou, while State Grid controls the grids in the rest of the nation.

The 2002 disintegration of the State Power Corporation had far-reaching impacts on China's power sectors, leading to the long-term monopolistic advantages of the two grid companies and relatively weak positions of the Big Five power generation companies. Since almost all the transmission and distribution networks are in the control of the two grid giants, they naturally have superior bargaining power over the Big Five which altogether supply only half of the power needed by China to the grids. Ironically, the Big Five, with traditional businesses all focusing on thermal power generation, are also major participants in renewable energy production, especially in the wind power sector. Guodian, Huaneng, Datang, and Huadian have been the top four wind project developers for years, accounting for about half of China's installed wind capacity. Only two private developers, namely China Wind Power and Tianrun, are among the top 15 wind power developers, accounting for about 3.3 percent of total installed wind capacity (Deutsche Bank Group 2012, pp. 10–11). China Three Gorges Corporation (CTGC), the centrally controlled hydropower grant, and provincial thermal SOEs such as Power Beijing and Ningxia Power, are also significant players in producing wind power in China.

CTGC, one of the top hydropower producers in the world, was established in September 1993 and was responsible for construction of the Three Gorges Dam, the world's largest hydroelectric power plant that went into operation in 2008. In 2002, the CTGC established the subsidiary company China Yangtze Power, which took over operations and management of the Three Gorges Dam and the smaller-sized Gezhouba Dam. Besides the construction of the Three Gorges Dam, the central government also authorized CTGC to develop other gigantic hydroelectric projects in the mainstream and tributaries of the upper reaches of the Yangtze River, including the massive hydropower plants at Xiluodu, Xiangjiaba, Wudongde, and Baihetan. As the operator of the Three Gorges and Gezhouba hydropower projects, China Yangtze Power sells its electricity via China State Grid Corporation mainly to central China (Hubei, Hunan, Henan, Jiangxi, and Chongqing), east China (Shanghai, Jiangsu, Zhejiang, and Anhui) and Guangdong province.

This wind market dominance by SOEs focusing on conventional energy development could mean that Chinese wind power plants are prone to yield to pressures from state-owned thermal/hydro plants or grid companies in the overall picture of power overcapacity. Despite their increasing investment and involvement in wind power generation, electricity SOEs such as the Guodian Group, the Huaneng Group, and CTGC remain preoccupied with their bulky thermal or hydropower businesses, and which distracts them from making full use of the wind turbines under their management.

China's solar PV power generation is similarly dominated by SOEs focusing on conventional energy. Despite the growing role of private developers such as the Golden Concord Group, the State Power Investment Group owns an installed capacity of 11.66 GW for solar power generation, making it the largest solar power generator not only in China but also in the world at large (State Power Investment Group 2018). Of its power generation portfolio, both wind and solar shares are growing very fast, but they are still dwarfed by the Group's staggering capacity in thermal power and hydropower, which stands at 74 GW and 22 GW, respectively (State Power Investment Group 2018). Of the Group's total installed capacity, thermal and hydropower account for 58.7 percent and 17.5 percent, while solar and wind only account for 9.3 percent and 11 percent, respectively. As China's major renewable energy player, the State Power Investment Group has been facing internal competition from various energy sectors, which could hamper its fast-growing renewable capacity from being efficiently utilized.

GOVERNMENTAL REGULATORS VERSUS STATE-OWNED ENERGY GIANTS

The major Chinese energy giants are centrally managed SOEs directly administered by the State-Owned Assets Supervision and Administration

Commission (SASAC) of the State Council. The executives of these energy corporations and other SOEs that are directly subordinate to the SASAC form a powerful elite group within the CPC framework. Around 50 managers who are deeply embedded in the CPC system of personnel patronage are appointed directly by the CPC Central Organization Department: they have ministry-level standing, and their appointments are finally reviewed and approved by the Politburo Standing Committee. The other 100 or so top managers are appointed by SASAC itself, and while their status is not quite exalted, they are also important people (Naughton 2008, p. 2). Such a political ranking system in favor of traditional energy producers in the fields of fossil fuels, hydropower, or even nuclear power generation places the emerging renewable energy developers in an uneven playing field for market competition.

Besides their long-existing political connections with top leadership, the fast-expanding profitability and assets of conventional energy giants in the sectors of fossil fuels and hydropower have dramatically reinforced their economic and political clout in the bureaucracy. These companies were already influential interest groups in the command economy era and this has only increased due to the vast amounts of wealth generated since then. Generally speaking, all SOEs have become more capital intensive. Between 1998 and 2010, the amount of assets per employee in SOEs rose by nearly seven-fold in nominal terms, from RMB 200,000 (US$31,700) to RMB 1.37 million (US$217,000), at a considerably faster rate of increase than those in domestic private firms (from RMB 92,000 to RMB 307,000 per employee) and foreign-funded enterprises (from RMB 275,000 to RMB 555,000 per employee) (Tong and Huang 2012). Among all the centrally and locally managed SOEs, the energy companies – especially the NOCs – make the largest share of profits. Despite such surging profitability, most state-owned energy giants, including the three NOCs, are not required to turn over a significant part of their profits to their putative owners, the Chinese government in the name of 1.3 billion Chinese citizens.

The specific issue of how to handle company dividends, or after-tax profits, dates back to 1994, when the central government, as it was implementing tax reform, made a rather casual decision that SOEs would no longer remit after-tax profits to the government (Naughton 2008, p. 1). After more than a decade of high-speed expansion, many centrally run SOEs have become so economically and politically powerful that their vested interests become the main obstacles that prevent the central government from collecting substantial dividends from their after-tax profits. Although the SASAC has been trying for years to force the NOCs and other super-rich SOEs to pay higher dividends to the government, the largest shareholder of these SOEs, it only achieved limited success with petroleum, electrical, and telecommunication enterprises finally agreeing to turn over a maximum of 10 percent profits to the treasury.

The rise of state-owned energy giants is closely connected with the transformation of the nation's regulatory structure and the corporatization process of SOEs. Many of the energy regulatory administrations in the command economy era were transformed into gigantic SOEs themselves, but the newly established watchdogs such as the SASAC and the State Electricity Regulatory Commission were not powerful enough to fill the vacancy and supervise those ministerial-level SOEs. The Ministry of Petroleum was converted into the CNPC in 1988 to handle petroleum activities in China, and the other two NOCs also originated within government departments. The establishment of the ministerial-level State Power Corporation (SPC) in 1997 was also perceived as a sort of preparation to fill in the vacuum left over by the dismantling of the Ministry of Electric Power one year later. In contrast, instead of being a full-fledged ministry like the powerful Ministry of Finance or the Ministry of Commerce, the SASAC is only defined as a special entity under the supervision of the State Council (*guowuyuan teshe jigou*). Its responsibility is to supervise the preservation and increment of the value of the state-owned assets through statistics and auditing, and to manage wages and remuneration of the top executives of the supervised SOEs.

Given the trend of separating enterprises from the government (*zhengqi fenkai*), a serious problem occurred when the central SOEs were forming a partnership to resist the regulatory authority. This complicated regulation pertaining to central–local government relations and the state–business nexus (Tsai 2011, p. 522). The government's effort to solve the problem of unifying government and state-run businesses through disbanding industrial ministries in the 1980s and 1990s led to the new issue of SOE monopoly or oligopoly in the absence of an efficient regulatory authority. In the power industry, between 1997 and 2002, the SPC not only acted as a monopoly company through the vertical integration of generation and transmission but also a corporatized ministry, because most officials in the defunct Ministry of Electric Power had transferred to the SPC after 1998. The first general manager of SPC, Shi Dazhen, was the last minister of Electric Power (Tsai 2011, p. 530). Even though the central government established the State Electricity Regulatory Commission (SERC) one year after the splitting of the SPC into two grid companies and five power generation corporations in 2002 to regulate the national power sector, the lack of specific regulatory authority remained unchanged since the SERC was not a cabinet ministry as the Ministry of Electric Power was inside the State Council. SERC thus does not have sufficient administrative power to supervise gigantic SOEs such as the ministerial-level State Grid Corporation and the China Three Gorges Corporation.

Being devoid of full discretion and sufficient resources, SERC is only able to play a coordinative role in establishing a unified system and formulating national plans and rules for the whole electricity industry comprising thermal power, hydropower, nuclear power, and wind power. More importantly, SERC has been sharing monitoring authority with other government entities like the NDRC, as national energy planners, and SASAC, as controlling shareholders of those power companies. For instance, the NDRC holds exclusive authority over reviewing and approving (*shenpi*) power projects and setting prices, while the SERC performs partial administrative and regulatory duties (Tsai 2011, p. 534). The performance of top executives in the giant state-owned electricity enterprises is evaluated by SASAC and sometimes by the CPC's Central Organization Department based on corporate profitability, tax revenues, and personal political loyalties, which has made electricity SOE leaders often ignore the instructions of SERC.

Just like many other Chinese regulatory bodies, SERC and its regional branches always suffered from the problems of understaffing and insufficient financial resources. According to the circular on electricity sector reform issued by the State Council, SERC was equipped with only 98 regular staff, which includes one chair and four vice chairs (General Office of the State Council 2003). In contrast, the State Grid Corporation alone had more than 1.58 million employees and China Southern Power Grid hired about 130,000 people. The understaffing problem existed even in the omnipotent NDRC, with the nickname of 'small State Council', whose Energy Bureau once had only about 30 employees before its upgrade to the vice-ministerial level NEA in 2008. Besides the NDRC, SASAC, and SERC, the Ministry of Finance, Ministry of Ecological Environment, Ministry of Natural Resources, Ministry of Science and Technology, Ministry of Industry and Information Technology, Ministry of Housing and Urban-Rural Development, and Ministry of Commerce are currently involved in the energy-related regulatory process (see Figure 8.1).

A peculiar political culture is that all SOE chief executives are bestowed administrative ranks (*xingzheng jibie*), which has made energy policy-making even more decentralized and dispersed. The general managers of China's major state-owned energy companies enjoy the ministerial-level or vice-ministerial-level rank that is equivalent to, if not higher than, many senior regulators, which enables them to initiate autonomous energy projects and policies and sometimes advance corporate interests at the cost of national ones. The head of the State Grid Corporation is a ministerial-level (*zhengbuji*) cadre as is the NDRC Commissioner, SASAC Commissioner, and SERC Chairman, so it becomes a mission impossible for these regulators to block the vested interests of State Grid Corporation from influencing policy-making in the power sector. Turf wars among various energy institutions occur frequently, leading to confusion in energy law-making and inefficient implementation of these laws and regulations. Since the high-profile governmental reorganization in 2008, the NEA under the NDRC has been endowed with more political clout than its predecessors, but due to its vice-ministerial rank, it is still not powerful enough to mediate the bureaucratic infighting that disrupts energy policy-making.

In the concept of 'fragmented authoritarianism' identified by Kenneth Lieberthal in China's decision-making (Lieberthal 1992), all institutions and all individuals of any importance hold a specific administrative rank, and this rank is a critical determinant in the geometry of policy bargaining. In the days of the planned economy, the large state industries were either ministries or affiliated bureaus, hence, partly for this reason, the few remaining large SOEs still retain considerable influence at the highest levels of government (Shirk 1993, pp. 107–15). The conventional energy SOEs in the sectors of fossil fuels and hydropower have proved adept at defining the energy policy agenda, and regulatory agencies like the NEA and the SERC more often than not find themselves in a disadvantageous position when bargaining with those industrial giants on the formulation and implementation of energy policies and plans, especially those supporting renewables. There has been widespread recognition among Chinese officials and energy experts of the need to make the country's energy institutions more effective by establishing a ministry of energy. Nevertheless, the status quo is still being favored by the energy establishment, and state-owned energy firms are reluctant to accept another political manager that would limit their direct access to China's leadership (Downs 2008b, p. 45). As China lacks an authoritative regulatory body to resolve disputes among powerful energy giants, their competing and conflicting interests are likely to undermine not only policy formulation and implementation, but also the drafting of energy laws and regulations.

China often faces the problem of connecting its impressive renewable power capacity to the national transmission power grid. This is largely due to the advantageous position that oligarchic grid companies have when bargaining with numerous renewable power plants scattered in interior areas where not all power generated can be transmitted. The conflict between the grid companies and electricity generation sectors (*wangdian zhizheng*) has been entrenched since the breakdown of the SPC into two grid companies and five power generation corporations in 2002, which led to a huge efficiency loss on the supply side and low prices of electricity sold by power plants to grid companies.

Besides the turf wars at the central level, the involvement of local governments deepens the fragmentation and decentralization in China's energy policy-making. Provincial, municipal, and county-level energy policy-making and implementation have been dispersed among various departments and bureaus. This includes the Local Development and Reform Commissions (branches of the NDRC), local subsidiaries of SASAC, finance, commerce, and environmental protection bureaus at various levels. In most cases, it is the local government, not the higher levels of these functional apparatus, that provides local environmental/technological/industrial agencies with their annual budgetary funds, approves institutional promotions, and determines increases in personnel and even allocation of resources such as cars, office buildings, and employee housing. Local leaders, including party bosses and administrative chiefs, all treat economic growth, fiscal revenue, and employment as policy priorities and are unwilling to see the local economy and investment hampered by implementation of strict energy and environmental regulations from the top.

High-speed economic growth is not only an important index of local officials' performance and linked to their future promotion prospects, it also increases local fiscal revenues and benefits local officialdom materially as a whole. All provincial party secretaries and governors enjoy the same (or even higher if the party secretary is a Politburo member) administrative ranking as the NDRC and SASAC Commissioner, so it has been very difficult for the central energy authority to coordinate energy policies among the provinces and supervise the implementation of relevant regulations and dictates. Since the early 1990s, China's officialdom has formed a culture whereby gross domestic product (GDP) growth is a major indicator of the performance of administrative chiefs and even determines their future promotions. Officials at all levels have been aware of the seriousness of China's energy shortage and low efficiency for a long time, but local administrations' common goal for GDP growth greatly weakened central government's coordination efforts for high efficiency and balanced development. Despite the NDRC's frequent warning against overinvestment and overcapacity in both the wind power generation sector and wind turbine manufacturing, generous subsidies from the Ministry of Finance and local governments' relentless pursuit of GDP led to massive expansion of China's wind power capacity and related component manufacturing and turbine assembling sectors, with signs of overheating in just a few years.

In reality, nevertheless, local authorities have not been concerned about the utility of these newly installed renewable facilities, largely because of pressures from conventional power producers and grid companies in the context of excessive local power supply. As the province with the richest wind resources in China, Inner Mongolia's installed wind power capacity approached 11.4 GW by March 2011, accounting for at least 30 percent of the nation's total; however, a large portion of the electricity generated there could not be transmitted elsewhere, which has caused huge losses to local windmills. Numerous wind turbines in Inner Mongolia were installed in remote areas too far away from the transmission grid and energy-intensive urban areas. At least 30 percent of installed wind capacity there was lying idle and not connected to the power grid (Forbes 2009).

POLICY PRIORITIES, INTEREST GROUPS, AND THE BUREAUCRACY

In China's five-layer vertical governance structure (central, provincial, municipal, county, and township), usually only interest groups at the central level are able to exert direct influence on central government policy-making. Policy priorities advocated by these central-level interest groups openly or implicitly thus have a significant impact on the policy-making process, with the political weights of these interest groups and interactions among them being the key variables in shaping final policy outcomes. For China's renewable energy policy, a part of the country's overall energy policy that overlaps its industrial and environmental policies, central policy-makers are positioned at the heart of the political games played by state-level stakeholders and interest groups in relevant sectors.

Before reform and opening up, the CPC traditionally emphasized the unity of the fundamental interests of all people in China and denied the diversity of interests. The reform initiated by Deng Xiaoping has created revolutionary changes in China, and with the expansion of the non-state economic and social sectors, new social stratums with different interests have emerged (Yang 2007, p. 1). At the state-level, various interest groups were derived from the fragmentation of the authoritarian bureaucracy (Lieberthal 1992, pp. 1–30) in the reform era, when the model of state capitalism was engaging in hybrid strategies that attempt to incorporate a state-led developmental path into a partially marketized economy. In such a state-centric economic model, all parochial interests not only originated from certain bureaucratic sectors, but were also still keeping close ties with the state apparatus on a regular basis through which they could seek to influence policy-making or implementation in favor of their own interests.

As mentioned in the introduction, China's energy policy-making process, is to a large extent, vertically organized in what is called energy system (*nengyuan xitong*), with party-state leaders at the top making final decisions on specific energy issues, including those about renewable energy development. However, the dependence on others for implementing China's energy policy has reduced these leaders' personal influences and magnified the institutional and pluralistic impacts on the process. Like in other issue areas, Chinese renewable energy politics, albeit being vertically organized from central to various local levels, has been constantly compromised by horizontal conflicts with formidable institutional players in such conventional energy systems as coal, petroleum and natural gas, hydropower and nuclear power systems or other non-energy systems including environmental, commerce, and foreign policy systems.

There are several actors in China's energy bureaucracy that have a strong voice in the energy policy-making process. The NEC is a key coordination entity in this regard. It is an inter-agency mechanism headed by the Premier and composed of 20-plus ministers and commissioners. The conventional players that play a pivotal role in its affairs include the NDRC and the NEA affiliated to it. Apart from these, there are five key institutional actors whose voices the CPC cannot neglect: large energy production SOEs such as the Big Five power companies, the three NOCs, major coal, hydropower, and nuclear power companies, the State Grid Corporation and other related ministries and departments, such as the China Electricity Council (CEC) and the Ministry of Ecological Environment. CEC, a joint social organization of China's power enterprises, functions as a bridge between the government and power enterprises, serves its members by appealing to the government on their behalf, and protecting their legal rights, and encourages its members to fulfill their social responsibilities, and also promotes the coordinated development of the whole industry (China Electricity Council 2018).

Among all the institutional pressures on energy policy formulation, those from oil, coal, and electricity-transfer business juggernauts are beginning to become highly influential within the government. This is a major cause for concern as these organizations are highly driven by their own commercial interests. China's renewable energy market from its beginning has been fragmented and decentralized, making most renewable developers unable to compete with conventional energy oligarchs in relation to influencing policy. Government departments and grid dispatchers with vested coal interests have been accused of exerting administrative power to disrupt wind and solar power production. The symbiotic relationship between coal-fired plants and grid companies exacerbates the transmission bottleneck faced by low-carbon alternatives. Like traffic congestion on a highway, overproduced electricity supply from varied resources including stable coal-fired thermal plants and intermittent wind and solar power plants results in electrical jams on the grid. In winter, grid operators often instruct wind and solar power plants to curtail wind turbines and solar panels in order to make transmission capacity available for 'must run' coal-fired units so they can continue operation to satisfy thermal heating obligations (Deutsche Bank Group 2012, p. 33)

As stakeholders of local thermal power plants, local authorities in many cases tend to help these companies bypass strict energy directives and curb competition from local renewable players. When designing renewable energy policies, Chinese policy-makers expect local governments to cooperate in enforcing those policies, laws, and regulations. In practice, however, this high degree of administrative cohesion does not exist in places where local government interests diverge sharply from those of the central planners and create substantial obstacles to strict enforcement of both national and local energy policies. As Chubb (1983, p. 9) points out, the most salient feature of contemporary energy politics is the battle being fought against the cooperative mode of policy-making, and against the public and private organizations that carried it out. This rule applies not only to democracies, but also to fragmented authoritarianism that practices state capitalism.

REFERENCES

Andrews-Speed, P. (2004), *Energy Policy and Regulation in the People's Republic of China*, The Hague: Kluwer Law International.

Bachman, D. (1998), 'Structure and process in the making of Chinese foreign policy', in S. Kim (ed.), *China and the World: Chinese Foreign Policy Faces the New Millennium*, Boulder, CO: Westview Press, pp. 34–54.

Chen, G. (2009), *Politics of China's Environmental Protection: Problems and Progress*, Singapore: World Scientific.

China Electricity Council (2018), 'CEC in brief', accessed 10 April 2018 at http://english.cec.org.cn/No.86.index.htm.

China Shenhua (2012), 'Company Introduction', accessed 2 April 2015 at http://www.csec.com/htmlen/orgAbout/page_1_1_1.htmlShenhua.

Chubb, J. E. (1983), *Interest Groups and the Bureaucracy: The Politics of Energy*, Stanford: Stanford University Press.

CNBC News (2018), 'China set to top Japan as world's biggest natural gas importer', 3 January 2018, accessed 1 April 2018 at https://www.cnbc.com/2018/01/03/china-set-to-top-japan-as-worlds-biggest-natural-gas-importer.html.

Deutsche Bank Group (2012), *Scaling Wind and Solar Power in China: Building the Grid to Meet Targets*, Deutsche Bank Group report, accessed 1 March 2018 at file:///C:/Users/eaicg/Downloads/Scaling%20Wind%20and%20Solar%20Power%20in%20China_Building%20the%20Grid%20to%20Meet%20Targets.pdf.

Downs, E. S. (2008a), 'Business interest groups in Chinese politics: The case of the oil companies', in C. Li (ed.), *China's Changing Political Landscape: Prospects for Democracy*, Washington, DC: Brookings Institution Press, pp. 121–41.

Downs, E. S. (2008b), 'China's "new" energy administration', *China Business Review*, November–December, 42–5.

Forbes (2009), 'Weaknesses in Chinese wind power', 20 July 2009, accessed 5 January 2015 at www.forbes.com/2009/07/20/china-wind-power-business-energy-china.html.

General Office of the State Council (2003), *Circular of the General Office of the State Council Concerning the Organizational Structure of State Electricity Regulatory Commission* [*guowuyuan bangongting guanyu yinfa guojia dianli jianguan weiyuanhui zhineng peizhi neishejigou he renyuan bianzhi guiding de tongzhi*], accessed 1 May 2018 at http://www.gov.cn/zhengce/content/2008-03/28/content_7690.htm.

Lieberthal, K. (1992), 'Introduction: the "fragmented authoritarianism" model and its limitations', in K. Lieberthal (ed.), *Bureaucracy, Politics, and Decision Making in Post-Mao China*, Berkeley, CA: University of California Press, pp. 1–30.

Naughton, B. (2008), 'SASAC and rising corporate power in China', *China Leadership Monitor*, 24, accessed 1 May 2018 at http://media.hoover.org/sites/default/files/documents/CLM24BN.pdf.

Shirk, S. (1993), *The Political Logic of Economic Reform in China*, Berkeley, CA: University of California Press.

State Power Investment Group (2018), 'About us' [*jituan jieshao*], accessed 1 March 2018 at http://www.spic.com.cn/jtgk_dhzy/jtjs/.

The Straits Times (2018), 'China sets up new ministries to oversee environment, natural resources', 13 March 2018, accessed 2 April 2018 at http://www.straitstimes.com/asia/east-asia/china-sets-up-new-ministries-to-oversee-environment-natural-resources.

Tong, S. and Y. Huang (2012), *China's State-Owned Enterprises in the Post-Crisis Era: Development and Dilemma*, Background Brief No. 694, Singapore: East Asian Institute (EAI).

Tsai, C.-M. (2011), 'The reform paradox and regulatory dilemma in China's electricity industry', *Asian Survey*, 51 (3), 520–39.

Tu, J. (2011), *Industrial Organization of the Chinese Coal Industry*, Working Paper 103, Program on Energy and Sustainable Development (PESD), Stanford University, accessed 2 March 2018 at https://carnegieendowment.org/files/China_Coal_Value_Chain_Kevin_Tu3.pdf.

Yang, G. (2007), *Interest Groups in China's Politics and Governance*, Background Brief No. 361, Singapore: East Asian Institute (EAI).

6. Disputes over how to go low-carbon

The Chinese central government's renewable energy ambition has been motivated not only by its concerns over energy shortage, but also by mitigation imperatives in the backdrop of climate change and smog pollution. China's favorable international image in the early stage of international climate negotiations has been changed dramatically and adversely since the country's total energy consumption and carbon emissions started to grow rapidly after China's accession to the World Trade Organization (WTO) at the end of 2001. As the world's largest carbon emitter, China, before the Copenhagen Climate Summit in 2009, was blamed by international media for wrecking the last-ditch talks and lacking sincerity in its intentions of working out a binding treaty on emission cuttings (*The Guardian* 2009; Spiegel Online International 2010). Therefore, since the very beginning, China's renewable energy development has been pushed by its mitigation commitment amidst intense international pressures.

Mitigation imperatives coupled with energy shortage has pushed China to foster the aggressive low-carbon electricity strategy that has embraced hydro, nuclear, wind, and solar power as major solutions against the backdrop of climate change. Hydropower and nuclear power, despite being more environmentally controversial than other renewables such as wind, solar, and geothermal energy, have been perceived by China as clean alternatives to conventional fossil fuels in a low-carbon-electricity scenario. With a historical lag in the country's hydro and nuclear power sectors, China has been planning aggressively for these two low-carbon alternatives, which are vying for supportive policies, state investment, and market shares vis-à-vis wind and solar power in the overcapacity era.

FROM ENERGY INTENSITY TO CARBON INTENSITY TARGETS: CHANGE OF POLICY PRIORITY BOOSTS LOW-CARBON SECTOR

As the world's largest carbon emitter, China contributed about 30 percent of the global total carbon dioxide (CO_2) emissions from the burning of fossil fuels and some industrial processes in 2014, 15 percentage points higher than the emissions share of the United States (Boden et al. 2017), the second largest emitter, which had been notorious for its high-carbon lifestyle for decades. Moreover, in the first decade of the twenty-first century, China had more than doubled its carbon emissions (US Carbon Dioxide Information Analysis Center 2011); if such a trend continues, most of the mitigation effect contributed by industrialized nations through cutting their emissions compulsorily might be neutralized by the newly added greenhouse gas emissions from China alone. The logic that size and growth rate do matter in terms of global emissions control has made China one of the most conspicuous nations under scrutiny of the international community, with fears that the unchecked emissions increase of the emerging superpower might jeopardize the fair play of global actions against climate change.

China, in an effort to alleviate international pressure and strengthen its soft power in climate politics, has been making more pledges in clean energy and emission reduction. Through proposing carbon intensity reduction targets on top of the long-pursuing energy intensity goals, China showed a certain flexibility and shrewdness in coping with the newly emerged challenges. Actually energy intensity, referring to energy consumption per unit of gross domestic product (GDP), is positively correlated to the variation of carbon intensity, which measures carbon emissions per unit of GDP. Both as efficiency indicators, energy intensity and carbon intensity can be brought down by technological changes as well as structural changes in the economy, for example, a move from heavy industry to a service economy (Lovejoy and Lee 2005, p. 379). Yet the two indicators will move differently when the country is replacing fossil fuels with low carbon alternatives such as hydropower, nuclear, wind, and solar power, without any significant improvement in energy efficiency. On that condition, it can still achieve a substantial reduction in carbon intensity, but not in energy intensity.

Despite China's satisfactory performance in the field of energy efficiency between 1980 and 2000 when its GDP quadrupled and energy consumption only doubled, which meant about 50 percent drop in energy intensity, the country has been frustrated by the rebound of energy intensity since 2001 when heavy and chemical industries started to grow rapidly in China. China had set an aggressive goal of cutting its energy intensity by 20 percent in its 11th Five-Year Plan period (2006–10). It seemed to be a mission impossible in the first two years, when it only managed to cut energy consumption per unit GDP by 1.23 percent in 2006 and 3.27 percent in 2007 (Chen 2010, pp. 56–7). On the eve of the Copenhagen Climate Summit in 2009, China, for the first time, declared that it was targeting a hefty 40–45 percent cut in carbon intensity, the amount of CO_2 emitted per dollar of GDP, by 2020 compared with a 2005 baseline. The main limitation of an intensity-based target is that, although it can lower an emission growth trajectory below the projected business-as-usual level, it is unlikely to result in an absolute decrease in emissions. A carbon-intensity target is especially attractive to China and other developing countries that seek a transition to a low-carbon economy without taking on quantifiable emission-reduction commitments. The application of low-carbon energy solutions on a large scale could help China to reduce carbon intensity by large margins even without mandatory emissions cutting obligations or remarkable energy efficiency improvement.

China later on included the goal of cutting carbon intensity by 17 percent by 2015 as a binding target in its 12th Five-Year Plan (2011–15), which was passed by the National People's Congress, the top legislature, in 2011. This made its legal position equivalent to many industrialized nations' mitigation commitments under the Kyoto Protocol, which had been transformed into domestic legislation. During Chinese President Xi Jinping's state visit to the United States between 22 and 28 September 2015, the two countries issued a joint presidential statement on climate change, in which Xi promised to lower carbon intensity by 60–65 percent by 2030 based on 2005 level. The Chinese side also agreed to implement a 'green dispatch' system to favor low carbon sources in the electric grid in China, and to launch a national emission trading system in 2017.

As China's GDP continues to grow rapidly, its total carbon emissions will also grow over the next decade even if carbon intensity has dropped significantly from the 2005 baseline; and the faster the economic growth is, the more CO_2 the country will emit. However, the carbon intensity targets, if strictly enforced, will help the largest carbon emitter to avert the 'no effort' scenario, and is therefore in line with the call from the Intergovernmental Panel on Climate Change (IPCC) for the substantial reduction of developing country emissions below business-as-usual (BAU) levels. Many analysts have agreed that a higher GDP growth rate would make the target easier to reach, simply because growth adds new, more efficient, and less carbon-intensive infrastructure, and the higher the growth rate, the larger the share of new and energy-efficient capital equipment in the economy as a whole (Chandler and Wang 2009, p. 3; Seligsohn and Levin 2010, p. 5).

In reality, China's economy growth experienced significant slowdown of below 8 percent since 2012, a low-growth threshold embraced by Chinese President Xi Jinping as the 'new normal' (Wong 2014). Such a geared-down GDP growth has made both the energy intensity and carbon intensity targets more difficult to achieve, as room for increase of energy consumption or carbon emissions was constrained by decelerated GDP figures. Under such a 'new normal' circumstance, the proposed carbon intensity targets became more stringent for China to achieve, requiring the country to develop low-carbon energies in a more aggressive manner. The same carbon intensity targets would allow China to increase total emissions by 30–40 percent over the 15-year period from the base year, if the annual growth rate is maintained at 6 percent, whereas they could increase by 75–90 percent in a scenario of 8 percent economic growth per year (Seligsohn and Levin 2010, p. 3).

The mitigation imperatives in an economic 'new normal' scenario has forced the central government to stimulate the low-carbon energy sector in a more urgent manner. The Chinese government has been scrutinizing the relationship between economic growth and energy utilization not only from an emissions perspective, but also from an energy security perspective (Lewis 2007, p. 156). Bringing down carbon intensity and improving energy efficiency are actually the main elements of China's long-pursuing 'no regret' strategy, which implies actions providing fringe benefits to the country regardless of whether the threat of global warming is real. Climate change is not only an environmental issue for China, but also a good opportunity to address the country's serious energy shortage problem caused by growing demand, inefficient use, and limited energy reserves. Energy policies have dominated China's development strategy for a long time and now the government exploits global concern about climate change as a golden opportunity to improve energy efficiency and boost clean energy production. Driven by such a 'no regret' strategy, China's low-carbon energy sector started to go full steam ahead across the board.

COMPETITION WITHIN LOW-CARBON ELECTRICITY SECTOR

As in many other countries, the electric power sector is vital in order for China to run its modern economy, with fossil fuels providing the bulk of the energy needed to generate electricity. Now, with new carbon intensity targets pledged by the Chinese leaders in the international arena, hydropower and nuclear power, together with renewables such as wind, solar, and geothermal energy, have been boosted with full speed in a low-carbon-electricity scenario. From 1991 to 2013, the proportion of hydro, nuclear, and wind power in China's total energy production rose from 4.7 to 10.9 percent, while that in China's total energy consumption grew from 4.8 to 9.8 percent (National Bureau of Statistics of China 2014, p. 72). The proportion of low-carbon electricity investment in China's total investment in the power industry scaled up from 29 percent in 2005 to 75 percent in 2013 (Mathews 2014).

China has been formulating the world's most ambitious policies with enormous investment in place to foster the aggressive construction of new hydropower and nuclear plants, in addition to its impressive input in such emerging renewable energies as wind and solar power. Between 2012 and 2017, all the four low-carbon power sectors experienced fast development, with the shares of wind, nuclear, and solar in total installed capacity edging up substantially, mainly at the cost of thermal power and hydropower. By the end of 2012, China's installed hydropower capacity was 248.9 GW, accounting for 22 percent of its total installed capacity in power generation that stood at 1,144.9 GW, while on-grid wind power reached 60.8 GW and made up 5 percent, nuclear 12.6 GW (1 percent) and solar power merely 3.3 GW (0.3 percent) (China Electricity Council 2013). Five years later, by the end of 2017, China's installed hydropower capacity increased to 341.2 GW, accounting for 19 percent of China's total installed capacity of 1,777 GW, while on-grid wind power reached 163.7 GW and made up 9 percent, nuclear 35.8 GW (2 percent) and solar power 130.3 GW (8 percent) (Figure 6.1) (China Electricity Council 2018).

Considering the exorbitant costs and risks involved in developing low-carbon energy, China's great leap forward in promoting non-fossil fuels has been closely related to the government's robust support for and heavy subsidies to the emerging low-carbon electricity industry. Both hydro and nuclear power projects are environmentally risky and need considerable front-loaded investment. Wind and solar power projects are capital-intensive as well, despite negligible fuel costs. In addition, wind and solar power generation can be highly variable at different timescales, and therefore pose substantial challenges to incorporating the electrical generation into grid systems, which could in turn increase costs for energy demand management, load-shedding, storage solutions and system interconnection. In promoting the production of low-carbon alternatives, the Chinese government has taken a differentiated approach to support policies, with variations in targeted growth scopes in mid- and long-term plans, which is profoundly changing the electricity generation market's existing structure and the country's energy mix dominated by coal burning.

However, as all these four non-fossil fuel sectors are heavily reliant on the government's supportive policies, the competition among themselves for favorable policy targets and fiscal supports has been extremely intense besides their contests with coal-fired thermal power in China's electricity market. Although non-fossil fuels are on track to become winners as a whole vis-à-vis thermal power, they will still witness broadly

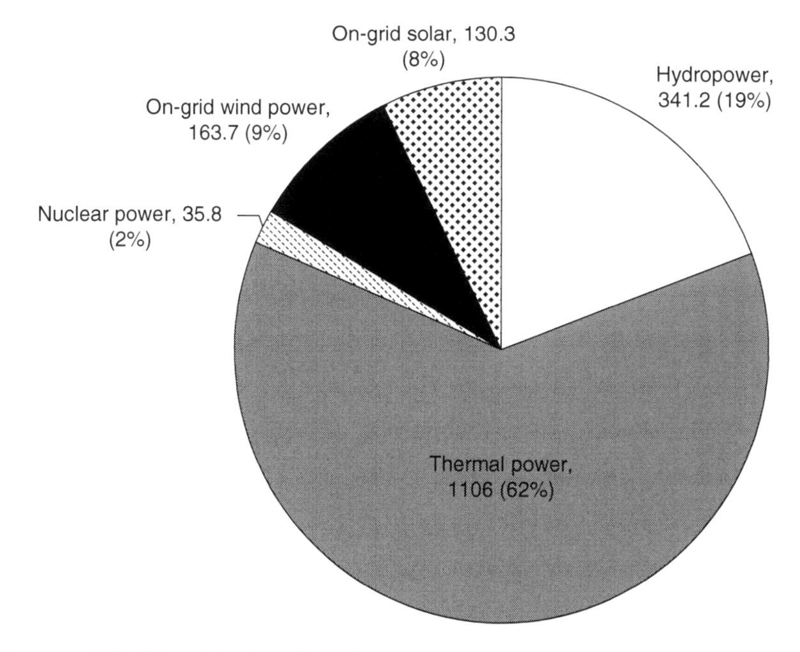

Source: China Electricity Council 2018.

Figure 6.1 Existing installed capacity of various electricity sectors in China by end of 2017 (in GW)

different growth trajectories among them, which will result in significant redistribution of their proportions in China's power market. Compared with other democracies, the authoritarian one-party regime in China faces much less resistance from civil society when pushing forward those hydropower and nuclear mammoths that may pose potentially serious threats to the local ecology. As an attractive solution to address climate change and energy shortage due to their negligible carbon emissions and low operating costs, hydroelectricity projects have long been favored by China, which has the largest number of big dams and the largest installed hydropower capacity in the world. Haunting concerns over the negative ecological and social impact of massive dams once made the government more cautious about hydropower and thus slowed down the construction of such projects.

Interest groups benefiting from the construction and operation of China's hydropower stations, however, have been well connected to politically powerful figures such as the ex-Premier Li Peng and thus are able to lobby energy administration for resumed support after temporary slowdown in the early 2000s. Li, a conservative political figure who began his career in a hydropower station located in northeast China, served as the vice-minister and minister of power between 1979 and 1983. Supported by other conservative Party elders, he subsequently joined the Politburo Standing Committee, the top echelon of power of the Communist Party of China (CPC), in 1987 and retained his seat till 2002. Li's experience as a hydropower engineer and a minister in charge of China's power industry established his leading position in the bureaucracy representing the interests of pro-hydropower groups. As a firm advocate of building hydropower projects along the Yangtze River, including the environmentally controversial Three Gorges Dam, Li paid 15 official visits to the site of the Three Gorges to examine the feasibility and construction of the dam between 1982 and 2002 (Li 2003). Besides Li, another top Chinese leader, Jiang Zemin, also played a vital role in promoting gigantic hydropower projects. Jiang had presided over all the major decision-making concerning the Three Gorges project since 1989, when he became the CPC General Secretary (Li 2003, p. 12). Li perceived hydropower as a 'clean and reliable' renewable energy for the twenty-first century, suggesting using the profit from the operation of Three Gorges Project to develop hydropower resources in the upstream of Yangtze River (Li 2003, p. 14). Such a rollover of profit into future development would inevitably snowball the investment in gigantic hydropower stations, leading to excess investment and overcapacity, and heightened environmental risks.

With studying the experience of the former Soviet Union, Li was also deeply involved in China's nuclear power generation, paying at least eight inspection tours to the Daya Bay Nuclear Power Plant located in coastal Guangdong province and bargaining with foreign leaders on deals of importing nuclear power generation facilities (Li 2005, pp. 1094–196). Li recalled his nuclear odyssey in his book *From Start-up to Development: Li Peng's Diary on Nuclear Power*, a memoir published in 2004 that offered a detailed account of his involvement in the country's nuclear development (*South China Morning Post* 2004). He mentions in the book China's authoritarian manner of suppressing strong opposition from the Hong Kong public in 1986, when the Chinese government was pushing ahead with the Daya Bay project at the doorstep of Hong Kong. Faced with strong objections in Hong Kong, Beijing sought to ease public fears by persuading anti-nuclear petitioners of the safety of nuclear energy while refusing to budge on the decision. Li said that the experience had taught him that stability in Hong Kong must be maintained, and that the mainland could not be 'jostled' by petitioners (*South China Morning Post* 2004). Such a hardline approach of handing NIMBY ('not in my back yard') movements against environmentally risky thermal power, hydropower, or nuclear power plants has fostered the leapfrogging development of these controversial energy projects in China.

During Li's tenure, China's wind and solar power development was still in the infancy stage and did not draw much attention from him. With his impressive policy legacy and formidable lingering influence in China's power industry, Li's preference over hydropower and nuclear power has been deeply embedded in the energy administration, with his protégés and children being placed in key positions of power development. His son, Li Xiaopeng, had been chief executive of China Huaneng Group, one of the 'Big Five' state-owned power generation companies, for many years before he was named vice-governor of the coal-rich Shanxi province in 2008. Li Xiaolin, Li Xiaopeng's sister, had been leading the state-owned China Power International when she was transferred in 2015 to serve as vice-president of China Datang Corporation, another company among the 'Big Five'.

Despite Li and his pro-hydropower/nuclear power faction's advocacy, the environmental impact of these low-carbon power plants has been a growing concern, affecting the government's decision to approve large-scale projects. The construction of the Three Gorges hydropower complex had provoked a long-lasting debate over the pros and cons of this largest-ever hydropower project, and even today, the controversy continues unabated in China. In 1992, when the then Chinese Premier Li Peng submitted a proposal for the Three Gorges Dam to the National People's Congress, which is usually seen as pliant, there was strong opposition against the proposal and was ultimately passed with the smallest margin in the legislature's history.

China passed the Law on Evaluation of Environmental Effects in 2002, requiring stakeholders of all power projects constructed within its territory to submit a comprehensive evaluation of environmental impacts on the ecosystem before construction begins (PRC Law on Evaluation of Environmental Effects 2002, Articles 3 and 4). The law, which gives state environmental regulators veto power over the approval of all kinds of new projects, including energy projects, encourages stakeholders, specialists, and the public to participate in the evaluation of environmental effects (PRC Law on Evaluation of Environmental Effects 2002, Article 5). Along with increasing public awareness of the negative environmental effects associated with large-scale projects, especially gigantic hydropower stations, the ratification of the Law on the Environmental Impact Assessment led to a slowdown in hydropower expansion in subsequent state plans. Whether the sudden slowdown of hydropower projects was related to Li Peng's complete retirement in 2003 was still unknown. In 2003 and 2004, proposals to build a string of dams on southwest Nu

River (also known as Salween) sparked a fierce debate in China. Diverse reporting perspectives and heated debates in the mass media therefore shaped public opinion, which in turn influenced government policies. Consequently, in April 2004, the State Council halted the construction of the hydropower project that was planned on the Nu River.

Previous radiation accidents that happened elsewhere also raised safety concerns about nuclear power, and this explains why China had lagged behind in the nuclear power sector. Such anxiety dissipated in the early 2000s as energy demand soared and China gained confidence in operating its own nuclear power plants. Nevertheless, China suspended nuclear plant construction nationwide for 20 months after the onset of the Fukushima Daiichi nuclear disaster in Japan in 2011, in response to public concerns about nuclear safety. Although the government resumed nuclear plant construction along the eastern coastal area in late 2012, most of the planned nuclear projects located in the hinterland, which may be susceptible to much greater disastrous consequences such as radioactive leakages, continued to be suspended for a longer period.

The year 2011, the beginning of the 12th Five-Year Plan (2011–15), witnessed the government's renewed enthusiasm for large hydropower projects, with enormous funds for investment in the construction of environmentally controversial dams and an ad hoc Energy Five-Year Program to formally prioritize the development of hydropower among various forms of power generation. Based on Li Peng's rollover plan of using profit from the Three Gorges Project for reinvestment in upstream hydropower stations, China started to invest 400 billion yuan (US$62 billion) in 2011 in the construction of four huge hydroelectric dams (Xiluodu, Baihetan, Wudongde, and Xiangjiaba) on the Jinsha River, a tributary of the Yangtze River (*China Daily* 2011). The installed capacity of the four dams totaled 43 GW, which almost doubled that of the Three Gorges Project, the world's largest power station and one of the most environmentally controversial projects in modern China.

Meanwhile, after several years of suspension of approval of nuclear power stations due to the Fukushima Daiichi nuclear disaster, the Chinese government decided to speed up the construction of nuclear reactors to achieve a targeted installed capacity of at least 58 GW by 2020 (World Nuclear Association 2018). As nuclear plants can be built close to centers of demand along the coastlines, which are distant from many wind, solar, and hydro sites located in western China, pro-nuclear interest groups are gaining grounds against the other three low-carbon energies, especially in the context of high curtailment for renewables. State-owned enterprises (SOEs) have always played dominant roles in the construction and operation of nuclear power plants, with the China Guangdong Nuclear Power

Group (CGNPC) and China National Nuclear Corporation (CNNC) having full control of the market. Since 2007, the government has diversified the ownership structure of nuclear power. The state-owned China Power Investment Corporation (CPIC) was given ownership of nuclear power plants along with CGNPC and CNNC. Due to safety and technology concerns, other state-owned thermal power producers, such as Huaneng, Datang, Huadian, and Guodian, have been allowed to participate in nuclear power projects but cannot hold majority stakes.

Foreign direct investment (FDI) has been encouraged in the construction and management of nuclear power plants on the condition that Chinese partners hold the majority of shares while foreign enterprises need to possess advanced technology. With regard to the manufacture of equipment, FDI had been encouraged in the production of nuclear power reactors of over 600 MW between 2004 and 2007. However, after the Chinese Ministry of Commerce revised the 'Catalogue for the Guidance of Foreign Investment Industries' in 2007, FDI has only been encouraged in the manufacture of equipment for nuclear power plants with a larger capacity of over 1 GW (Yang and Pan 2011, pp. 5–6). China is heavily dependent on foreign technology in the design of nuclear power reactors, but it has been vigorously pursuing indigenous technology over which it has intellectual property rights. China's two mainstream reactor types were based on the second-generation-plus design from Areva, France, and the third-generation design from Westinghouse, a US company. The CNNC has been working on the China Experimental Fast Reactor, which is the key to fourth-generation technology. It made a breakthrough in successfully testing the first China Experimental Fast Reactor in July 2010, making China the eighth country in the world to own such technology (*Xinhua News* 2010). Compared with the third-generation reactors, which had a utility rate of uranium of just 1 percent, the China Experimental Fast Reactor reduced the consumption of uranium and had a utility rate of more than 60 percent. The technological breakthrough in the fourth-generation reactor may, in the long run, help China to cope with the scarcity of natural uranium resources for nuclear power development.

China's prioritization of certain low-carbon electricity technologies is by and large related to the country's inherent geographic and meteorological advantages in tapping energy resources. Its land generally reduces in elevation from west to east, and its sloping topography causes humid sea air currents to penetrate deep into China's internal regions, thus bringing about abundant runoffs to the large rivers that flow eastwards. Due to these natural factors, China is leading the world in hydropower potential, most of which remains untapped in the less developed and populated southwestern and western regions. China's major river basins are also well within the monsoon zone of the Pacific, with over 50 percent of annual precipitation in most areas concentrated in the four months from June to September. Occasionally, July and August are the two months that receive more than 70 percent of regional rainfall. However, two-thirds of China's land is threatened by frequent flooding, and its coastal areas are ravaged by tropical typhoons seven times a year on average; its hinterland is also not spared from droughts, which occur almost every year during the dry seasons. The frequent incidence of devastating floods hence justified the Chinese government's construction of gigantic dams.

The fact that China's bountiful hydroelectric resources are mostly located in the southwest, whereas its wind and solar resources are concentrated in the vast but sparsely populated northwestern and northern areas, exacerbates the situation of imbalanced distribution of energy resources, as the economically vibrant eastern (coastal) area has high energy demand but is constantly afflicted with energy shortages. The most important area in China that is rich in wind resources covers the long inland belt that extends from Xinjiang, Gansu, and Ningxia provinces in northwestern China to Inner Mongolia and Hebei province in the north. Inner Mongolia alone accounts for about one-third of China's total installed wind power capacity. China's northwestern region of Tibet, Xinjiang, and Qinghai and Gansu provinces have the strongest solar radiation that is conducive to solar energy development, but the southeastern provinces, where China's economic powerhouse is, receive a much lower amount of solar radiation per square meter. The overconcentration of conventional coal, hydroelectricity, and emerging wind and solar resources in the western part of China has exacerbated the heated competition among various energy sectors, imposing enormous pressure on the country's transmission networks. China hence initiated the launch of the West–East Electricity Transmission Project, aiming to transmit power from the electricity-rich western regions to the economically prosperous eastern areas.

In the backdrop of overcapacity, various interior provinces prioritize certain types of power supply from their own provinces, often ignoring the central government's coordinative directives that ask them to use renewable energy transferred from other provinces. Hydropower-rich southwestern provinces such as Sichuan and Yunnan are not enthusiastic about developing solar or wind power. Northwestern provinces such as Xinjiang and Gansu are obsessed with wind and solar power development, while northern provinces such as Inner Mongolia and Shanxi are facing fierce internal competition among coal-fired thermal, wind, and solar, due to their abundance in all these energy resources. Other provinces located in central China, including Hubei, Hunan, and Jiangxi, are reluctant to use power transmitted from western provinces, as they themselves are facing internal power oversupply and want to sell electricity eastward to coastal provinces.

Benefiting from the imbalanced geographic distribution, nuclear power plants and offshore wind farms, which can be built close to centers of demand and thus save long-distance power transmission costs, have unique topographical leverages over other non-fossil fuels. Coastal provinces have been trying to alter the long-lasting overdependence on energy supply from hinterland areas through boosting their own low-carbon power industry. That, however, will inevitably intensify the competition among different energy sectors and various provinces, which will rely more on their lobbying activities with central authorities.

DIVERSE ON-GRID TARIFFS

China's power industry, inspired by the experiences of industrialized countries in power industry and electricity pricing reform, has become increasingly competitive since the early 2000s as a result of government-initiated structural reforms aimed at increasing competition and efficiency. In December 2002, the State Council ended the State Power Corporation's monopoly over the power industry by dismantling the corporation into 11 smaller companies, namely two electric power grid operators, five electric power generation companies, and four consultant and construction companies. The 'Big Five' generators, namely China Huaneng, China Datang, China Huadian, Guodian Power, and China Power Investment, contribute only about half of China's power output today, while the remaining half of the output is contributed by a large number of fragmented power generation companies generating different energy types including low-carbon sources. Other influential players at the national level include China Three Gorges Corporation (hydropower) and China National Nuclear Corporation (nuclear), and all these centrally administered power generation companies not only have to compete among themselves for bigger market share, but also need to vie with a large number of provincial power companies including Beijing Jingneng Thermal Power Company and Shenergy Company in Shanghai.

The disintegration of the State Power Corporation has had a far-reaching impact on China's power sectors, whereby the two grid companies, State Grid Corporation of China and China Southern Power Grid Company, hold long-term monopolistic advantage, and most power generation companies are relegated to relatively weak positions. Since almost all the transmission and distribution networks are in the control of the two grid giants, they naturally have superior bargaining power over all kinds of power generators, whose profit margins have been primarily determined by the 'on-grid electricity tariff' (*shangwang dianjia*), or the sales price between power generation companies and grid companies (Ma 2011, p. 2633). Despite progress in reforming electricity price mechanisms, including trials of setting tariffs through competitive bidding processes, China's on-grid electricity tariffs are still strictly regulated by the government. To foster the development of renewable energy resources, differentiated on-grid electricity tariffs have been adopted to provide incentives for renewable power producers including wind and solar.

Considering the comparable disadvantages of generating electricity from renewable sources, the PRC Renewable Energy Law (2005) addressed the core issues of pricing and fee-sharing for on-grid renewable energy through a government-set or government-guided scheme of feed-in tariffs. With the stipulation that 'grid power price of renewable energy power generation projects shall be determined by the price authorities of the State Council' (PRC Renewable Energy Law 2005, Article 19), the feed-in tariff imperative requires the following: that a fixed amount, adjusted according to renewable energy types and geographic locations, be added to the selling prices from renewable electricity generators to the grid companies, on top of the benchmark on-grid electricity tariffs of coal-fired power plants. As a financial incentive, the feed-in tariff allows the renewable generator to achieve positive returns on its investment, despite the higher production cost compared with conventional fuels. When Chinese policy-makers found that wind power, compared with solar energy production, was at a more mature stage at the time to be widely commercialized with modest government subsidies given, they decided to promote wind power vigorously as the main renewable resource because subsidizing solar generation was deemed too costly and inefficient. In 2006, the cost of electricity generated from solar power was about 3 yuan/kWh, while that of a typical coal-fired power plant was only around 0.22 yuan/kWh, and that of a wind power plant averaged at about 0.6 yuan/kWh (Li 2007).

Under China's differentiated on-grid pricing system, which essentially maintains the cost-accounting approach, nuclear power plants are rewarded with higher on-grid tariffs than coal-burning stations, most of which follow the benchmark prices that are higher than hydropower tariffs, while wind power prices are even higher than nuclear, and solar power prices are the highest. In 2009, the on-grid thermal tariff in China averaged about 0.35 yuan/kWh – higher than the hydro tariff averaging at 0.26 yuan/kWh, but lower than nuclear price ranging between 0.39 and 0.49 yuan/kWh – and wind price from 0.51 to 0.61 yuan/kWh. There was no unified on-grid tariff for solar PV until 2011, when the government

Table 6.1 China's regional wind power tariffs (2009 and 2016)

Region	Wind Resource	Tariff in 2009	Tariff in 2016
Resource Region I	Rich	0.51 Yuan/kWh	0.40 Yuan/kWh
Resource Region II	Modest	0.54 Yuan/kWh	0.45 Yuan/kWh
Resource Region III	Modest	0.58 Yuan/kWh	0.49 Yuan/kWh
Resource Region IV	Low	0.61 Yuan/kWh	0.57 Yuan/kWh

Notes:
Resource Region I (*yilei ziyuanqu*): Urumqi, Yili, Changji, Karamay, and Shihezi in Xinjiang Autonomous Region, and most parts of the Inner Mongolia Autonomous Region excluding Chifeng, Tongliao, Xing'an, and Hulunbeier.
Resource Region II (*erlei ziyuanqu*): Chifeng, Tongliao, Xing'an, and Hulunbeier of the Inner Mongolia Autonomous Region, Zhangjiakou and Chengde in Hebei province and Zhangye, Jiayuguan, and Jiuquan in Gansu province.
Resource Region III (*sanlei ziyuanqu*): Baicheng and Songyuan in Jilin Province, most parts of Xinjiang excluding Urumqi, Yili, Changji Karamay and Shihezi, the whole Ningxia Autonomous Region, most parts of Gansu province, excluding Zhangye, Jiayuguan, and Jiuquan, and part of Heilongjiang province.
Resource Region IV (*silei ziyuanqu*): the rest of China excluding the first three resource regions.

Source: NDRC (2009 and 2016), 'National on-grid wind power tariff list' [*quanguo lushang fengli fadian biaogan shangwang dianjia biao*] accessed 6 January 2013 and 27 April 2018, respectively, at www.sdpc.gov.cn/zcfb/zcfbtz/2009tz/W020090727530432780298.pdf and http://www.ndrc.gov.cn/zwfwzx/zfdj/jggg/201612/W020161228343602191833.pdf.

formally announced a feed-in tariff of 1 yuan/ kWh for solar photovoltaics (PV) projects scheduled to be completed after 2011 and another tariff of 1.15 yuan/kWh for projects completed before the end of 2011 (National Development and Reform Commission of China 2011).

Currently, on-grid tariffs for wind (Table 6.1), solar PV, and even nuclear are on a downward trajectory due to the sharp decline in related power production costs, while the government is considering the possibility of raising on-grid rates for hydropower or thermal projects to rebalance the competition in favor of conventional power sectors. Since 2008, the Chinese government has adopted a set of preferential pricing schemes to encourage wind power generation. The tariff for wind power was set by the NDRC in 2009 and classified into four levels to give differentiated support to projects located in different regions that vary in local wind resources (Table 6.1). In December 2016, the NDRC lowered the on-grid wind power tariffs across the board based on decreasing power generation cost and oversupply situation, with tariffs dropping 10–20 percent from the 2009 levels in all the four resource regions (Table 6.1).

Prior to 2009, when the solar power market was small, the approved feed-in tariff rate ranged between 4 yuan/kWh and 9 yuan/kWh based on the different characteristics of individual projects, about 10–20 times that of coal power. The exorbitant cost of producing on-grid solar power, together with the enormous cost disparities among projects in various localities, meant that the government had refused to adopt the fixed feed-in tariff, which policy-makers thought would lead to overcapacity and costly production in some regions. In 2010, grid companies received government subsidies of 5.3 billion yuan for purchasing power from wind power generators, but only received 63 million yuan and 83 million yuan in subsidies, respectively, when buying electricity from hydro and solar power plants (China's State Electricity Regulatory Commission 2011, p. 10). In certain localities, it was only after successful feed-in tariff and other subsidy programs benefiting regional solar power projects that the central government finally decided to grant nation-level support. A bidding process for a 10 MW program in the Dunhuang region of western China's Gansu province resulted in an on-grid price of 1.09 yuan/kWh in 2009. Such a price level was applicable then to local projects in surrounding regions with similar solar resources and equipment.

A year later, in April 2010, the NDRC announced that four solar power stations in western Ningxia Autonomous Region would adopt a new price level of 1.15 yuan/kWh, higher than that applied to the Dunhuang program. In 2011, the Chinese government introduced a nationwide feed-in tariff of 1 yuan/kWh, which offered above-the-market-price contracts for the generation of solar electricity. Against the backdrop of exacerbated solar PV overcapacity and the launch of anti-dumping and countervailing duty investigations by the United States and European Union (EU) regarding China-produced PV products, the Chinese central government planned to give an additional nationwide subsidy of 0.45 yuan/kWh of electricity generated by solar PV equipment, while many provincial governments decided to offer additional subsidies of 0.25–0.3 yuan/kWh, with a unified term of 20 years. The China Renewable Energy Industries Association projected that solar PV power generation could reach grid parity as early as 2014 (China Renewable Energy Industries Association 2011), a point at which solar PV can generate electricity at a levelized cost that is less than or equal to the price of purchasing power from the electricity grid, and thus become a contender for widespread development vis-à-vis conventional thermal power. According to this pathway to grid parity, if the on-grid solar tariff decreases from the 2011 level of 1 yuan/kWh at an annual rate of 8 percent while the thermal on-grid tariff rises 6 percent per annum, then grid parity would be realized in 2014 when the price of solar electricity from the grid is lower than coal-fired electricity.

The NDRC institutionalized the nationwide subsidy to all solar PV power plants by announcing a subsidy level of 0.42 yuan/kWh in a circular released in August 2013 (National Development and Reform Commission of China 2013), a 20 percent increase from the planned level of 0.35 yuan/kWh, which was proposed in its draft version half a year earlier. With such a higher-than-expected government subsidy granted to domestic solar power plants, China witnessed its PV installed capacity grow exponentially in the following years, having achieved a stellar performance similar to that of its wind power market in 2005–10. Ranked fourth in global solar power generation in 2013, China surpassed Germany as the country with the most installed PV power capacity by 2015 (*Bloomberg News* 2017).

Driven by the central authorities' policy of prioritizing non-fossil fuels in the energy mix, China has been experiencing a frenzy of low-carbon energy across-the-board since the beginning of the twenty-first century. Renewables in strict sense like solar and wind, as well as controversial options of hydropower and nuclear, have all been promoted by the government forcefully in order to bring down the proportions of carbon-intensive fossil fuels, especially coal, in the energy mix. China vowed to cut 2015 levels of carbon intensity by 18 percent by the end of 2020, part of a longer-term pledge to bring total greenhouse gas emissions to a peak by 2030. China's coal-dominant energy structure puts the country in an awkward position in the global climate change context. China has been one of the world's largest energy producers for many years, with three-quarters of its domestic energy supply coming from coal production. With verified coal reserves ranking third in the world after the United States and Russia, China's annual coal output and consumption is positioned number one in the world. Compared with its abundant verified coal reserves that accounted for 13 percent of the world's total, China's verified reserves of oil and natural gas, the other two major fossil fuels, are extremely limited. China's per capita average of coal resources is 50 percent of the world's average, while the per capita average of both oil and natural gas resources is only one-fifteenth of the world's average. Such an imbalanced structure of domestic energy reserves, which predetermines coal's dominant role in the country's energy mix, is the root of many energy/environment-related challenges faced by China today.

From the mitigation perspective, China's overall strategy of boosting the low-carbon power industry helps the country diversify its energy structure by relying on all sorts of non-fossil fuels, which will in the long run result in faster reduction of the carbon intensity index than energy intensity. In 2016, coal consumption dropped 4.7 percent and accounted for only 62 percent of China's total energy consumption, 2 percentage points lower than a year ago. At the same time, consumption of lower-carbon alternatives including hydro, nuclear, natural gas, and wind power accounted for 19.7 percent of total energy consumption, 1.7 percentage points up year on year (National Bureau of Statistics of China 2017).

China's recent achievement in reducing carbon intensity was in part related to the country's 'supply side' reform under President Xi Jinping's leadership, which had shut down numerous highly polluting thermal power plants and eliminated excessive capacities in steel, coal, and other heavy industries across the country. Yet such a 'supply side' reform has failed to address the overcapacity and excess investment in the low-carbon energy sectors, which, among themselves, are facing over-competition and low efficiency problems that have discounted China's mitigation effort. If China's low-carbon energy strategy is scrutinized from ecological, efficiency and safety perspectives, it will not be difficult to find out that massive state support, which to a large extent has ignored local demand conditions and distorted market competition, has produced such unintended consequences as adverse environmental and safety impact associated with hydropower and nuclear, and low efficiency of installed wind and solar capacity in interior provinces. When China's economy is slowing down into a 'new normal' stage, where its energy demand is also decelerating accordingly, rivalry among various non-fossil fuel sectors for stronger state support and bigger market shares is expected to intensify. China's energy consumption soared more than 200 percent between 1990 and 2010, but in recent years, such growth has significantly slowed down, with energy consumption increasing only 1.4 percent in 2016 (National Bureau of Statistics of China 2017). While the gradual stabilization of China's energy consumption is a good sign for global joint action against climate change, as it bodes well for the country to keep the promise of peaking its carbon emissions by 2030, such an energy 'new normal' could worsen the over-competition and overcapacity facing all non-fossil fuels, stifling the commercial viability of renewable energy projects in many localities.

The political connections between various low-carbon energy sectors and the Chinese government are not favorable towards renewables like wind and solar. Compared with major hydropower and nuclear power corporations, which are state-owned behemoths subject to state plans and government edicts, wind and solar power plants are mostly small and medium-sized generators with diverse ownership, and are therefore more sensitive to fluctuations in power pricing, production costs, and demand conditions. As discussed earlier, the 'Big Five' state-owned power companies, whose conventional businesses consist of coal-fired power generation, have been increasingly engaged in operating wind and solar power plants. Once these power companies are facing internal conflict of interest between their conventional (thermal) power businesses and new energy (renewable) businesses, in most cases, the nascent renewable sectors would have to give way to the more powerful coal-fired businesses. Besides, government plans, which are more precise in predicting future development trajectories of hydro and nuclear power, often lag behind in forecasting the proliferation rate of wind and solar power plants, which face higher supply and demand elasticity due to government subsidies, feed-in tariffs, and slumping production costs. Without effective participation of environmental NGOs or civic society that is sensitive about environmental impact of energy projects, renewables such as wind and solar do not have sufficient edge over environmentally controversial low-carbon alternatives in the context of power oversupply and transmission bottleneck.

REFERENCES

Bloomberg News (2017), 'China on pace for record solar-power installations', 20 November 2017, accessed 5 April 2018 at https://www.bloomberg.com/news/articles/2017–11–20/china-on-pace-for-record-solar-installations-as-forecasts-jump.

Boden, T. A., G. Marland, and R. J. Andres (2017), 'National CO_2 emissions from fossil-fuel burning, cement manufacture, and gas flaring: 1751–2014', Carbon Dioxide Information Analysis Center, Oak Ridge National Laboratory, US Department of Energy, accessed 2 January 2018 at http://cdiac.ess-dive.lbl.gov/trends/emis/top2014.tot.

Chandler, W. and Y. Wang (2009), 'Memo to Copenhagen: Commentary is misinformed – China's commitment is significant', web commentary from Carnegie Endowment for International Peace, accessed 5 January 2015 at www.carnegieendowment.org/files/Memo_to_Copenhagen_edits_Revised_12-14-091.pdf.

Chen, G. (2010), 'Carbon intensity: China's card for climate politics', *East Asian Policy*, 2 (2), 55–61.

China Daily (2011), 'China to prioritize hydropower', 23 June 2011, accessed 5 January 2017 at http://www.chinadaily.com.cn/business/2011-6/23/content_12763475.htm.

China Electricity Council (2013), 'Statistical brief on China's electricity industry in 2012' [*2012nian woguo dianli gongye gaikuang*], accessed 10 April 2018 at http://www.cec.org.cn/yaowenkuaidi/2013-02-22/97555.html.

China Electricity Council (2018), 'China's installed power capacity increases 7.6 per cent in 2017' [*2017 quanguo fadian zhuangji rongliang zengzhang 7.6%*], accessed 10 April 2018 at http://www.cec.org.cn/nengyuanyudianlitongji/hangyetongji/dianlixingyeshuju/2018-03-02/178238.html.

China Renewable Energy Industries Association (2011), 'China roadmap of photovoltaics development: A pathway to grid parity', accessed 6 November 2015 at http://www.creia.net/publish/report/121.html.

China's State Electricity Regulatory Commission (2011), *Power Industry Monitoring and Supervision Report 2010* [*dianli jianguan niandu baogao*], Beijing: State Electricity Regulatory Commission.

Lewis, J. (2007), 'China's strategic priorities in international climate change negotiations', *Washington Quarterly*, 31 (1), 155–74.

Li, L. (2007), 'China urges electricity suppliers to buy "green" power', accessed 5 January 2016 at http://www.renewableenergyworld.com/rea/news/article/2007/09/china-urges-electricity-uppliers-to-buy-greenpower-49879.
Li, P. (2003), *Diaries of Three Gorges Dams* [*Li Peng Sanxia Riji*], Beijing: Zhongguo Sanxia Chubanshe.
Li, P. (2005), *Diaries of Power Industry* [*Li Peng Dianli Riji*], 1–3, Beijing: Zhongguo Dianli Chubanshe.
Lovejoy, T. E. and J. H. Lee (2005), *Climate Change and Biodiversity*, Ann Arbor, MI: Yale University Press.
Ma, J. (2011), 'On-grid electricity tariffs in China: Development, reform and prospects', *Energy Policy*, 39, 2633–45.
Mathews, J. (2014), 'What does China's renewable energy revolution mean?', SPERI (Sheffield Political Economy Research Institute) Comment, accessed 5 October 2016 at http://speri.dept.shef.ac.uk/2014/08/20/chinas-renewable-energy revolution-mean/.
National Bureau of Statistics of China (2014), *China Statistical Abstract 2014*, Beijing: China Statistics Press.
National Bureau of Statistics of China (2017), *China's Statistical Bulletin on National Economy and Social Development in 2016* [*zhonghuarenmin-gongheguo 2016nian guomin jingji he shehuifazhan tongji gongbao*], accessed 2 April 2018 at http://www.stats.gov.cn/tjsj/zxfb/201702/t20170228_1467424.html.
National Development and Reform Commission of China (2011), 'The NDRC Circular on improvement of solar PV on-grid tariff policy' [*guojia fazhangaige-wei guanyu wanshan taiyangneng guangfu fadian shangwangdianjia zhengce de tongzhi*], accessed 7 January 2015 at http://www.gov.cn/zwgk/2011-08/01/content_1917358.htm.
National Development and Reform Commission of China (2013), 'NDRC's circular on the promotion of solar PV sector through price leverage' [*guojia fazhangaigewei guanyu fahui jiage ganggan zuoyong cujin guangfuchanye jian-kangfazhan de tongzhi*], accessed 9 November 2017 at http://www.ndrc.gov.cn/zwfwzx/zfdj/jggg/201308/t20130830_556127.html.
PRC Law on Evaluation of Environmental Effects (2002), accessed 2 April 2018 at http://www.npc.gov.cn/englishnpc/Law/2007-12/06/content_1382122.htm.
PRC Renewable Energy Law (2005), accessed 5 January 2018 at http://www.npc.gov.cn/englishnpc/Law/2007-12/13/content_1384096.htm.
Seligsohn, D. and K. Levin (2010), 'China's carbon intensity goal: A guide for the perplexed', China FAQs, accessed 5 January 2015 at www.chinafaqs.org/files/chinainfo/ChinaFAQs_China's_Carbon_Intensity_Goal_A_Guide_for_the_Perplexed_0.pdf.
South China Morning Post (2004), 'Li Peng recalls his nuclear odyssey', 5 October 2004, accessed 3 March 2018 at http://www.scmp.com/article/472972/li-peng-recalls-his-nuclear-odyssey.
Spiegel Online International (2010), 'How China and India sabotaged the UN Climate Summit', accessed 1 March 2015 at www.spiegel.de/international/world/0,1518,692861,00.html.
The Guardian (2009), 'How do I know China wrecked the Copenhagen deal? I was in the room', accessed 1 March 2015 at www.guardian.co.uk/environment/2009/dec/22/copenhagen-climate-change-mark-lynas.
US Carbon Dioxide Information Analysis Center (2011), 'Preliminary 2008–09

global and national estimates', accessed 3 January 2015 at http://cdiac.ornl.gov/ftp/trends/co2_emis/Preliminary_CO2_emissions_2009.xls.
Wong, J. (2014), *China's Economy 2014/15: Adjusting to the New Normal of Moderate Growth*, Background Brief No. 980, Singapore: East Asian Institute (EAI).
World Nuclear Association (2018), 'Nuclear power in China' (updated March 2018), accessed 2 April 2018 at http://www.world-nuclear.org/information-library/country-profiles/countries-a-f/china-nuclear-power.aspx.
Xinhua News (2010), 'China starts up first fourth generation nuclear reactor', 22 July 2010, accessed 5 May 2015 at http://news.xinhuanet.com/english2010/china/2010–07/22/c_13409085.htm.
Yang, M. and R. Pan (2011), *Nuclear Power Industry in China*, Background Brief No. 593, Singapore: East Asian Institute (EAI).

7. Five-Year Plans and energy policy priorities

As the world's leading practitioner of state capitalism with the heritage of a command economy, China has been using Five-Year Plans (FYPs) for decades to ensure predictable and controllable socioeconomic development along the trajectory preferred by the ruling Communist Party. Now officially called *gui-hua* (or program) instead of *ji-hua* (planning), China's FYP these days sounds more like an 'indicative' type of development planning, mainly outlining some strategic priorities in general terms without binding implementation targets. In reality, however, China's FYP for National Economic and Social Development is still essential to map out in five-year cycles major objectives and directions regarding the future development of economy, politics, society, culture, and environment.

Unsurprisingly, as one of the world's major energy producers and consumers, China always lists energy policy priorities in its FYPs. China's energy policy kit can be generally divided into two parts, that is, policies either related to energy supply or energy demand. Hence, the core of all energy-related policies by top planners always includes maximizing/diversifying energy supply and controlling/conserving energy demand. The scrutiny over policy priority changes in China's energy FYPs contributes to the understanding of how the central authorities have adjusted energy policy targets in response to new industrial and market conditions as well as concerns from different interest groups.

EVOLUTION OF FOSSIL FUEL POLICY PRIORITIES

China's energy supply policies cover four major areas, which include domestic production, import and transportation of fossil fuels (coal, oil and, natural gas), conventional power generation (thermal, hydropower, nuclear), power transmission, and finally, domestic supply from sources of renewable energy (wind, solar, biomass, geothermal, etc.). Self-sufficiency has been the recurring mantra of Chinese energy policy for decades, given that 90 percent of its energy demand relies on domestic resources largely due to its plentiful coal reserves. Since the country's energy consumption has been growing dramatically alongside its economic boom, the government has formulated a number of policies to enhance its coal supply capacity and regulate the industry for safety, environmental, and efficiency reasons. Coal production is the only fossil fuel sector in which private investment is allowed to enter without much restriction from the government. However, this resulted in the simultaneous increase in fatal accidents and pollution in privately owned small- and medium-sized coal mines. Due to this, the central government has been calling for the sustainable development of the industry.

The government has been encouraging mergers and acquisitions of large state-owned coal enterprises and minor ones to form conglomerates which would have a total annual production capability of more than 100 million tons. Miners in private or township enterprises who do not follow proper engineering methods have often ruined huge coal deposits, and many have lost their lives in fatal mining accidents. The central government decided to close township and village coal mines in the early 2000s, but this policy was obstructed by local officials who had stakes in lucrative mines. Subsequently, in 2012, the Chinese government raised the minimum requirement for safety investment in coal mining and urged mine operators to strengthen safety education for their workers. Coal mining enterprises must set aside a minimum safety investment of about US$5 per ton of coal output for mines having high levels of gas, and 19 US cents per ton for mines that are considered less dangerous. In contrast, the minimum safety investment required by previous regulations set in 2004 was only 4–10 US cents per ton of output for highly dangerous mines and 2–6 US cents per ton for mines considered lower risk.

Despite this, there were small- and medium-sized coal mines that did not conform to the reformed industrial policies. Those that continued to remain highly polluting and possessed poor safety conditions were closed down. In some instances, local officials were forced by disciplinary watchdogs to withdraw their investment and stakes in local coal mines. The huge profitability of China's coal mining sector relies on the country's soaring energy demand as well as the government's permission for coal price fluctuation based on supply–demand dynamics in the market. Unlike the oil, natural gas, and electricity prices that have been strictly regulated by the National Development and Reform Commission (NDRC), the prices of coal in different regions and for different purposes (power generation, residential heating, etc.) are left to float in the market.

The privatization of small- and medium-sized coal mines since the 1990s and marketization of coal sales have directly spawned numerous multimillionaires in coal-rich Shanxi, Shaanxi Province, and Inner Mongolia. The surging coal price squeezed the profitability of thermal powerhouses as they could not subsequently raise the selling price of electricity due to the government's fear of inflation and social instability. This conflict between the coal and power sectors finally became so severe that the NDRC had no choice but to alter the coal-electricity price linkage by allowing electricity prices to follow those of coal, to increase the profit margins the of power generation companies. In May 2011, the NDRC raised the electricity prices for industrial use for the first time since 2009 by an average 0.0167 yuan (0.25 US cents) in 15 regions to combat power shortages and losses of powerhouses. Despite this, due to government concerns over inflation, the electricity price for residents remained unchanged (*China Daily* 2011a).

When domestic coal supply became more and more expensive due to the country's insatiable energy demand, it also brought a window of opportunity for the development of other low-carbon energy alternatives, such as hydro, wind, and solar power. The central government, in fear of power production disruption and inflation upswing, started to encourage the import of coal from Australia, Indonesia, and Brazil while putting restrictions on the export of coal. Most of the coal producers and coal users in China do not have autonomy in the export or import of coal. Only the four giant state-owned enterprises (SOEs), namely the Shenhua Group, China National Coal Group, Shanxi Coal Import and Export Corporation, and China Minmetal, are authorized by the government to export some of their coal products. Those corporations that need to import coal from foreign countries need to obtain a license for running coal businesses and another one for importing coal from the local government. The year 2009 became the watershed for such coal market realignments when China, long a net coal exporter, suddenly imported a record 127 Mt. This was 63 times of the volume it imported in 2000 and nearly 15 per cent of all globally traded coal. In 2010, China emerged as the world's second-largest coal importer only after Japan and bought nearly 165 Mt on the back of booming demand and domestic transport constraints. Only one year later, China overtook Japan and became the world's largest coal importer (Cornot-Gandolphe 2014, p. 7).

The surge in coal demand is commercially driven due to the widening price gap between cheap international prices and relatively high domestic prices. In January 2014, the price difference between domestic thermal coal delivered from the northern port of Qinhuangdao to Southern China, and international shipments from Newcastle, Australia, or Indonesia expanded to more than US$20 per metric ton (Cornot-Gandolphe 2014, p. 9). Other motives have driven the increase in coal imports such as the need for environmental protection. In line with this aim, China's 12th FYP (2011–15) for the development of the coal industry formulated by the National Energy Administration pledged to encourage coal imports and cap annual domestic coal production at 3.9 billion tons by 2015. About 740 Mt of domestic coal production capacity would be newly added each year, most of which was for power generation (National Energy Administration of China 2012).

In this five-year period, China would continue to close down small- and medium-sized coal mines while supporting large and super-large coal enterprises, which were to account for 63 percent of China's total coal production capacity. Those medium-sized mines with more than 300,000 tons of annual coal production capacity were expected to contribute to 22 percent of the country's total capacity, while small-sized mines whose capacities were lower than 300,000 tons were to contribute only 15 percent. According to the FYP, by 2015, there would be ten coal enterprises whose annual coal production capacity exceed 100 million tons and another ten coal enterprises with annual capacity above 50 million tons (National Energy Administration of China 2012). In China's 12th FYP for coal industry development, the government tried to stabilize the coal output in the central area such as Shanxi and Henan Province while taking more effort to explore coal mines in the western part, such as Shaanxi, Yunnan, Guizhou, and Xinjiang. The government planned to encourage mergers and acquisitions among coal miners, thermal powerhouses, and coal transportation enterprises, with substantial support to be granted to the coal bed gas industry and carbon capture and storage (CCS) in the coal mines.

The surge in imported coal, triggered by widened price gap between domestic and imported coal and tightened environmental regulations, actually challenged China's energy strategy based on the principle of self-sufficiency. With the closedown of small- and medium-sized coal mines, China has to increase energy production from other domestic sources to ensure self-reliance. Although the country has been increasingly dependent on foreign oil, the Chinese government insisted on ensuring that its domestic crude oil output should remain at 200 million tons per year in the 12th FYP period. In comparison, during the previous FYP (2006–10), China's crude oil output had averaged 190 million tons annually. This indicated that its oil production capacity had reached its upper limit, leading to a deeper dependence on imports. In order to stabilize its domestic oil supply, efforts to exploit oil and natural gas, with the focus on major oil and gas basins such as Bohai Bay, Songliao, Tarim, and Ordos continued in the 12th FYP. China also attempted to explore new areas like the South China Sea area, over which it has sovereignty disputes with neighboring Southeast Asian countries.

Simultaneously, China aimed to ratchet up the construction of more national crude oil reserve bases in the 12th FYP. China proposed a program to establish strategic oil reserve bases in 2004 and took the leap in late 2007 when the NDRC announced the official establishment of a national oil reserve center to accelerate setting up of the oil reserves bases. The NDRC announced that the project would require three phases, over a total of 15 years, for completion. The four strategic oil reserve bases established in four coastal cities in the first phase, namely Dalian, Qingdao, Ningbo, and Zhoushan, have been operational since late 2008 with a total designed volume of 120 million cubic meters, equivalent to 140 million tons of oil. The second phase was due to be completed by 2012, with a total of 37.53 million tons of reserve volume (*China Daily* 2011b).

According to China's mid- and long-term oil reserve plan approved by the Chinese government, the country will complete the third phase of the project by 2020, establishing a total reserve volume equivalent to 100 days of imports, which may surpass 800 million tons based on estimates by analysts (*China Daily* 2011b). To further alleviate the energy shortage in the economically vibrant eastern areas along the coastlines, China sped up the construction of oil and natural gas pipeline networks, with another 350 billion yuan of investment in the 12th FYP. The plan had a goal of constructing oil and gas pipelines totaling 140,000 km by 2015, half of which were designed to transfer natural gas. The West–East Gas Pipeline project (*xi qi dong shu*) on such a scale will, in the long run, compete with the country's West–East Electricity Transfer project (*xi dian dong song*) that moves power from coal, hydro, wind, and solar sources eastwards through vast networks of electrical transmission lines.

One of the positive aspects of China's 12th FYP was the country's attention to the development of shale gas, coal-bed gas, tar sands, and other non-conventional energy resources, despite its technological shortcomings in these fields. The NDRC even formulated a specialized FYP for shale gas development in 2012. According to this document, all the three national oil companies (NOCs) (the China National Petroleum Corporation (CNPC), China Petrochemical Corporation (Sinopec) and China National Offshore Oil Corporation (CNOOC)) had started prospective work for shale gas production in regions such as Sichuan, Guizhou, and the Anhui Province. The plan set the goal of producing at least 6.5 billion cubic meters in 2015, with a total of 200 billion of exploitable shale gas reserves to be prospected across the country by that time (National Development and Reform Commission of China 2012). In the long run, China's shift to shale gas development may help the carbon-intensive economy to significantly reduce its discharge of carbon as a consequence of the coal-burning process. The NDRC estimated that if all the targeted shale gas outputs in 2015 were used in power generation, a total of 14 million tons of carbon dioxide emissions would be avoided yearly. The plan admitted that China still lacked core technology to make full use of the country's shale gas resources that are mainly present in the interior provinces such as Sichuan, Guizhou, Jiangxi, Yunnan, Jiangsu, Shaanxi, Henan, Anhui, Hunan, Hubei, Liaoning, and Xinjiang. The plan asserted that it was likely for the annual shale gas output to reach 60–100 billion cubic meters by 2020 (National Development and Reform Commission of China 2012).

FROM 12TH TO 13TH FYP: CHANGES OF LOW-CARBON TARGETS

Amidst increasing global concerns about climate change and energy security, the clean energy industry has emerged as one of the major business opportunities in China, which has been repeatedly addressed by various FYPs over the years. Considering the relatively high cost and risk involved in clean energy production, China's great leap forward in sharpening its competitive edge has been heavily dependent on the government's policy support based on targeted growth in the FYPs. Under the framework of the Renewable Energy Law and national energy targets, detailed institutional incentives and disincentives have been designed by various layers of the government to promote renewable energy development in China. In terms of promoting the production of solar and wind power, the government has been adjusting its policy prioritization from time to time, with variations in targeted growth scopes, subsidy amount, on-grid tariffs and other financial incentives.

By the end of the 11th FYP (2006–10), hydropower had been the focus of China's low-carbon energy strategy, contributing much more to the nation's electricity market than wind, nuclear and solar power combined. By the end of 2010, China's installed hydropower capacity was 213.4 GW, accounting for 22.2 percent of its total installed capacity in power generation that stood at 962.2 GW. Meanwhile, on-grid wind power and nuclear power only made up 3 percent and 1.1 percent of the total, respectively, and solar less than 0.1 percent (China Electricity Council 2011). During the 11th FYP period, the installed capacity of hydro and wind power grew 12.9 percent and 89.8 percent, respectively, per annum, while nuclear only achieved 9.6 percent annual growth, from a much lower baseline than the installed capacity of hydropower; the installed capacity of solar power, until 2010, remained at an insignificant level of 0.86 GW (State Council of China 2013). The year 2011, the beginning of the 12th FYP, witnessed the government's renewed enthusiasm for large hydropower projects, with enormous funds for investment in the construction of environmentally controversial dams and an ad hoc energy five-year program to formally prioritize the development of hydropower among various forms of power generation. As a step to change the stagnant development of hydropower and raise the share of non-fossil fuel to at least 15 percent in China's energy mix by 2020, the central government decided to add a staggering 120 GW in installed hydropower capacity from 2011 to 2015, or more than 50 percent of the existing total hydro capacity.

China's decision to resume the great leap forward in hydropower occurred in the wake of the Japanese Fukushima nuclear crisis in March 2011. As low-carbon alternatives to conventional fossil fuels, nuclear and hydropower have been competing with each other for years, with greater support to nuclear projects in the light of the controversies over the operating Three Gorges Dam before 2011. The nuclear disaster in Fukushima, however, dramatically shifted public opinion in favor of hydropower in China. With its large land mass and watersheds, China still has a huge potential for developing its economically and commercially feasible hydro resources, as compared with other advanced hydropower-rich countries such as Norway, Japan, and Canada, where more than 60 percent of the economically viable hydro resources have already been tapped. China has the largest economical hydro potential in the world, but only about one-third has been utilized.

China is bound to confront mounting pressure from both home and abroad when it accelerates its hydropower projects. This may result in massive displacements and might provoke landslides and earthquakes in addition to adverse ecological consequences in surrounding areas. The State Council has yielded to public concern over the problematic Three Gorges Dam by admitting unforeseeable urgent problems as a consequence of the world's biggest hydropower plant. Activists from Thailand, Laos, Cambodia, and Vietnam have blamed Chinese dams on the Mekong River for contributing to a drop in river flow and producing irreversible change to the river's ecology. The existence of strong pro-hydro interest groups and insatiable investment demand may mean that any hydro slowdown is temporary. Some suspended hydropower projects like the Jinanqiao Hydro Station on the Jinsha River have received approval to run from Beijing. Despite the government's open statement that the Three Gorges Dam has pressing geological, human, and ecological problems, there is no evidence that such concern would further decelerate hydro expansion in the next five to ten years, as it did in the previous five years. Contrary to the global trend that focuses on small hydro, China is still interested in mega-projects like the Xiluodu (12.6 GW) and Xiangjiaba (6 GW) hydropower stations on the Jinsha River.

In the middle of the 12th FYP period (2011–15), the Chinese government, in response to new industrial conditions of leapfrogging photovoltaics (PV) manufacturing capacity and nuclear power technology, as well as environmental concerns, adjusted its low-carbon electricity strategy, shifting the focus from hydro to wind, solar, and even nuclear power. If it was not for the strong opposition from the environmentally conscious civil society against massive dams, the share of hydropower in total installed capacity would have been much larger. Economic planners expected such a shift in policy priority to reverse the hydro-dominant trend, as the 12th FYP projected that solar, wind, and nuclear power would grow annually by 89.5 percent, 26.4 percent, and 29.9 percent, respectively, and hydropower by only 5.7 percent (State Council of China 2013). Emerging trends, driven by the more sophisticated energy markets, volatile energy production costs, reassessment of inherent geographical conditions and environmental impact, and increased support to domestic industries, were changing the value of non-fossil fuels. Since then, China has started to make concerted efforts to increase wind and solar power production to boost the share of renewables in national power consumption.

Besides fiscal subsidies, the government has adopted a set of preferential electricity pricing schemes to encourage renewable power generation (see Chapter 6). Since 2009, the tariffs for wind power have been set by the NDRC and classified into four levels to provide differentiated support to projects located in regions that vary in local wind resources. As the 12th FYP period experienced excessive investment and high curtailment in the wind power sector, central planners had to gradually cut fiscal subsidies towards wind power producers through substantially lowering the on-grid wind power tariffs across the country in December 2016 (see Table 6.1). The cut in subsidies and on-grid power tariffs for wind power reflected a gravity change in China's renewable strategy, from wind-first to prioritization of solar power in the 13th FYP period. When the solar power market was small in the 11th FYP period, the approved feed-in tariff rate ranged between 4 yuan/kWh and 9 yuan/kWh based on the different characteristics of individual projects, 10–20 times that of coal power. The exorbitant cost of producing on-grid solar power, together with enormous cost disparities among projects in various localities, prevented the central government from providing solar PV operators fiscal support comparable with that granted to wind power producers. In 2010, grid companies received government subsidies of 5.3 billion yuan for purchasing power from wind power generators, while they only received 63 million yuan and 83 million yuan in subsidies when buying electricity from solar power and hydropower plants, respectively (China's State Electricity Regulatory Commission 2011, p. 10).

The Chinese government in the first year of the 12th FYP formally announced a feed-in tariff of 1 yuan/kWh for solar PV projects to be completed after 2011 and another tariff of 1.15 yuan/kWh for projects completed before the end of 2011. Even with that, China's solar power industry was still wrestling with hurdles such as high costs, technical difficulties, and low efficiency. Against the backdrop of exacerbated solar PV overcapacity and the launch of antidumping and countervailing duty investigations by the United States and the European Union over China-produced PV products, the Chinese central government planned to give an additional nationwide subsidy of 0.45 yuan for each kWh of electricity generated by solar PV equipment, while many provincial governments decided to offer an additional subsidy of 0.25–0.3 yuan on top of that, with the unified term of 20 years. The NDRC institutionalized the nationwide subsidy to all solar PV power plants through announcing a subsidy level of 0.42 yuan/kWh in a circular released in August 2013 (National Development and Reform Commission of China 2013), a 20 percent increase from the planned level of 0.35 yuan/kWh which had been proposed in its draft version half a year ago. With such a higher-than-expected government subsidy granted to local solar power plants, China, then the world's fourth largest solar PV power producer, witnessed its PV installed capacity grow exponentially in the latter half of 12th FYP period.

The 12th FYP anticipated that solar would grow 89.5 percent annually, whereas wind's growth rate would drop to 26.4 percent. Compared with the goals set in the 12th FYP, the NDRC proposed the targets of increasing installed capacity of wind power by 63 percent to 210 GW, and that of PV power by 144 percent to 105 GW in the 13th FYP (2016–20) period (National Development and Reform Commission of China 2016). In terms of low-carbon electricity generation, hydropower by comparison has contributed much greater capacity to China's electricity market than wind, nuclear, and solar PV power combined. However, if the above targets are achieved by 2020, the total installed capacity of wind and PV power would be 315 GW, close to the expected hydropower capacity of 340 GW by 2020 (National Development and Reform Commission of China 2016).

A reverse trend has occurred during the 12th FYP, when solar grew 122 percent and wind only 33 percent annually (National Development and Reform Commission of China 2016). By the end of 2015, China's installed capacity of wind power and PV power generation reached 129 GW and 43 GW, both ranking first in the world and accounting for 7.8 percent and 2.6 percent, respectively, of China's total installed capacity (National Development and Reform Commission of China 2016). In 2015 alone, China's solar energy capacity increased by a massive 74 percent over the year before, with a modest gain of 34 percent in wind. At the same time, China was importing 30 percent less coal and consuming 3.7 percent less overall (National Bureau of Statistics of China 2016).

A significant but often-ignored component of China's renewable energy development is power generation from biomass sources. In the 12th FYP, China set the target of installing capacity equivalent to 13 GW for biomass power generation, but in reality, the total biomass capacity grew 13.4 percent yearly to 10.3 GW by 2015 (National Development and Reform Commission of China 2016), falling short of the target due to insufficient local support and distraction from the solar and wind power zest. The 13th FYP paid more attention to the efficiency issues regarding the utilization of installed wind and PV capacity, intermittence problem caused by renewable power fluctuations, transmission difficulty, and geographic uneven distributions of wind and PV power generation projects. To address the long-existing problem of high curtailment rates in wind and PV, the 13th FYP proposed for the first time the targets of annual power outputs for the wind and PV power generation as compared to the 12th FYP, which only set targets for the installed capacities of the two renewables. To promote sales of electricity from renewable sources, the 13th FYP set the goal of having wind power compete with local coal-fired power generation on the same price platform and PV power price to be equal to the local selling price from grid companies (National Development and Reform Commission of China 2016). The 13th FYP has realized the urgency of addressing the high curtailment rates of renewable energy projects, urging more wind and PV projects to be located in China's eastern and central provinces instead of resourceful but sparsely populated western provinces.

Currently, on-grid tariffs for wind, solar PV, and even nuclear are in a downward trajectory due to the sharp decline in related power production costs, while the government is considering the possibility of raising on-grid rates for hydropower or thermal projects to rebalance the competition in favor of conventional power sectors. Although the central government has decided to prioritize the development of renewables such as wind, solar, biomass, and geothermal in the 13th FYP, vested interests in benefiting from conventional energy sectors are still powerful and adept at lobbying for on-grid tariff adjustment favorable towards them. Despite the government's preference for hydropower construction, China's differentiated power pricing system is not in favor of operational hydro powerhouses. Coal-burning power plants are rewarded with higher grid feed-in tariffs than hydropower stations, while wind power prices are even higher and solar power prices the highest. If the government raises feed-in rates for hydropower projects in the near future, that would be a financial incentive to further boost hydro development.

In 2020 and 2030, the country's electricity consumption may hit 7 trillion kWh and 9 trillion kWh, respectively (State Information Center of China 2017). The rapid growth of China's power sector has made the country the largest market for power transmission and distribution and a major consumer of smart grid technology. A specialized FYP has been drafted by the Ministry of Science and Technology to address the expanding gap between the surging installed capacity scattered in different parts of the country and the insufficient transmission capacity of conventional grids. On the supply side, a massive built-up of gigantic hydropower stations and numerous renewable power stations would increase the demand for grid connections and efficient management of these intermittent power resources. In China's wind power expansion, only about one-third of these newly built windmills have been connected to the grid due to the shortage of grid networks and difficulty in managing intermittent currents. In the 12th FYP, the State Grid invested 286 billion yuan (US$45.4 billion) in smart grid construction to meet the need of connecting wind capacity of 90 GW and solar capacity of 8 GW to the grid (*China Daily* 2012), but this failed to have an immediate effect on solving the high curtailment issue.

Nuclear power involves nuclear fission reactions to generate electricity for civilian purposes. In comparison with other forms of renewable energy, it operates at nearly full capacity and incurs a lower cost, similar to or even lower than traditional coal-fired power generation. Nuclear power plants can be built close to centers of demand, thus saving long-distance power transmission costs. Nuclear power is important for China, especially in the coastal areas that are far away from the coal fields but where the vibrant economies are located. Nevertheless, nuclear power only accounted for 1 percent of China's total installed electricity generation capacity and 1.9 percent of total electricity generated in 2009. This was significantly lower than that of many other countries including most industrialized nations and some major emerging markets. As of July 2011, China's installed nuclear power capacity ranked only ninth in the world. Amidst the growing consciousness about climate change and energy shortages, China intends to speed up the construction of nuclear reactors in the near future, with targeted installed capacity reaching 58 GW by 2020, with 30 GW more under construction (World Nuclear Association 2018).

The country is rapidly becoming self-sufficient in reactor design and construction, as well as other aspects of the fuel cycle. In this regard, it revised its energy industry strategy from 'moderately develop nuclear power' in the 10th FYP (2001–05) to 'actively develop nuclear power' in the 11th FYP (2006–10). Nuclear power was also underlined in a 5 trillion yuan (US$0.78 trillion) alternative energy plan between 2011 and 2020 drafted by the National Energy Administration. According to the revised plan that supports surging nuclear development, the Chinese government reset its 2020 target of installed nuclear power capacity at 86 GW, more than double from its earlier target of 40 GW in the 'Nuclear Power

Medium- and Long-Term Development Plan' (2005–20) drafted by the NDRC in 2007 (Zheng and Mao 2009).

If the revised 2020 target was fulfilled, China's installed nuclear power capacity would see a nine-fold increase from the 9 GW capacity it had achieved by the end of 2009. In reality, however, due to safety concerns in the aftermath of the Fukushima nuclear disaster, the Chinese government's long-term target, as outlined in its Energy Development Strategy Action Plan 2014–20, was revised down to a capacity of 58 GW by 2020, with another 30 GW under construction (World Nuclear Association 2018). Previous plans for the establishment of a number of nuclear power plants in inland provinces such as Jiangxi, Anhui, Hunan, Hubei, Jilin, Henan, Sichuan, and Gansu have also been suspended. Even so, China today still has the world's largest nuclear power capacity under construction, both planned and proposed. As of April 2018, mainland China had 38 nuclear power reactors in operation, about 20 under construction, and more about to start construction (World Nuclear Association 2018).

TENSION BETWEEN EFFICIENCY AND SUFFICIENCY TARGETS

The government set the target of cutting energy intensity – an efficiency indicator referring to energy consumption per unit of GDP – by 20 percent from 2006 to 2010 in its 11th FYP. Since 2009, the year of the Copenhagen Climate Summit, China has added carbon intensity, another efficiency index measuring carbon emissions per unit of gross domestic product (GDP), into its energy planning system. China's 12th FYP set the dual targets of cutting energy intensity by 16 percent and carbon intensity by 17 percent between 2010 and 2015, while its 13th FYP aimed at further reduction by another 15 percent and 18 percent, respectively, by 2020. Behind China's more ambitious carbon intensity targets, as compared with its energy intensity goals, was the country's high-stakes gamble in expanding its low-carbon energy production capacity. In other words, China has been trying to improve the efficiency benchmark through lifting up low-carbon energy self-sufficiency rates and lowering fossil fuel proportions in the energy mix.

In order to achieve its goals, China used economic and taxation tools such as preferential purchase of renewable power, preferential taxation policies towards energy efficiency projects, favorable financing policies, tax-rebate cuts on high-energy-intensity exports, and compulsory government procurement of energy-saving products. Detailed energy-efficiency actions included petroleum substitution, surplus heat utilization, construction of energy-efficiency buildings, public transport development, energy-saving auto development, innovation of coal-fueled industrial boilers, and green lighting projects. Since 2007, the government started to use energy intensity index together with overall economic and investment figures to measure performances of local officials.

Some regional energy-efficiency and low-carbon projects like the state-level pilot zone in Hubei and Hunan Provinces and the Singapore-China Eco-City in Tianjin were launched. Since 2009, carbon intensity became another efficiency gauge that gave a big boost to low-carbon energy development. Finally, the country achieved an 18 percent decrease in energy intensity and a 20.8 percent decrease in carbon intensity during the 12th FYP period (State Information Center of China 2016), both of which overshot the original targets. The government estimated that the fulfilment of the 12th FYP targets had saved the equivalent of 670 million tons of coal for the five years.

The main limitation of an intensity-based target is that, although it can lower an energy and emission growth trajectory below the projected business-as-usual level, it is unlikely to result in an absolute decrease in energy consumption and carbon emissions. As China's GDP continues to grow rapidly, its total energy production and consumption will also grow over the next decade even if energy intensity drops significantly from the baseline. However, considering China's low per capita energy consumption and its exemption from legally binding emission-cutting targets, such a pledge is a strong one that could bring about considerable energy saving and emission reduction from the business-as-usual line calculated before additional mitigation measures are put in place. A carbon-intensity target is especially attractive to many developing countries that seek a transition to a low-carbon economy without taking on quantifiable emission-reduction commitments. The target, if strictly enforced, will alleviate international pressure on the largest energy consumer and carbon emitter in the world. This would put it in line with the call from the International Energy Agency (IEA) and Intergovernmental Panel on Climate Change (IPCC) for the substantial reduction of energy consumption and carbon emissions in developing countries below the business-as-usual scenario. The real impact of this type of intensity goal on China's total energy consumption will largely depend on the actual annual GDP growth rate to be fulfilled over the planned period.

Bringing down energy/carbon intensity and improving energy efficiency are actually the main elements of China's long-pursuing 'no regret' strategy, which implies actions providing fringe benefits to the country regardless of whether the threat of global warming is real. Such energy efficiency targets are essential for the country to tackle the serious energy shortage problem caused by growing demand, inefficient use, and limited energy reserves. China's parliament, the National People's Congress, enacted the Law on Conserving Energy in 1997. This was amended ten years later to align it with the global context of climate change. In the face of growing criticism over GDP-centrism, the amendment to the Law on Conserving Energy articulated for the first time that the state:

> will implement a system of accountability for energy conservation targets and a system for energy evaluation whereby the fulfilment of energy conservation targets is taken as one part of the evaluation of local people's governments and their responsible persons. (PRC Amended Law on Conserving Energy 2007, Article 6)

In addition, the amendment also promoted the energy conservation policy from the status of 'a long-term strategy for national economic development' in the original version to that of 'a basic policy of China' (PRC Amended Law on Conserving Energy 2007, Article 4), which is comparable with the policy of family planning that had been strictly implemented for more than 30 years. The amendment required local governments to formulate stricter energy-saving standards than those set by the central government.

The relationship between economic growth and energy utilization matters greatly both from an environmental perspective and an energy security perspective. Although China is still able to supply around 90 percent of the energy it needs for industrialization and urbanization, in the long run, energy scarcity could become a real bottleneck for China's sustainable development due to its increasing dependency on imported oil and even coal, and lower-than-average per capita energy resources. China's per capita average of both coal and hydropower resources is 50 per cent of the world's average, while the per capita average of both oil and natural gas resources is only one-fifteenth of the world's average. Energy policies have dominated China's development strategy for a long time and today the government is exploiting global concerns about climate change as a golden opportunity to improve energy efficiency and boost clean energy production. China's energy policy outcome is determined by a string of domestic and international considerations such as economic growth, energy security, perceived vulnerability, sovereignty, and international image, however, among all factors, the regime's preoccupation with meeting the voracious demand for energy arising from the swelling economy has set the tone of its overall energy strategy.

Nevertheless, this kind of symbiotic relationship between China's energy sufficiency and efficiency strategies could, in the long run, lead to overdependence on newly added capacities of low-carbon energies including hydropower, nuclear, wind, solar, and geothermal power, and overlook the real efficiency issues addressed in documents such as the PRC Law on Conserving Energy (1997) or FYPs. In practice, overdevelopment of low-carbon alternatives has resulted in fierce competition among different renewables, grid transmission deficit, geographic mismatch between supply and demand, and consequently, overcapacity and low efficiency in the non-fossil fuel sector.

Energy self-sufficiency has been the cardinal principle guiding China's energy policy since the establishment of the People's Republic of China in 1949, while the rule of efficiency was incorporated into China's energy policy-making at a much later stage, when the country started to encounter energy shortage and environmental challenges in the reform era. Since Chinese officials and entrepreneurs in the energy industry have been more adept at dealing with sufficiency issues rather than efficiency challenges, the promotion of low-carbon energy production naturally becomes one of the most convenient ways of improving the efficiency indicator of carbon intensity. In the fight against climate change, China proposed the targets of lifting up the proportion of non-fossil fuels in total energy consumption to 15 percent by 2020, and 20 percent by 2030. Large parts of the recent 12th and 13th FYPs have been increasingly devoted to the ambitious development of all low-carbon energies. Compared with the Law on Conserving Energy, the Renewable Energy Law has received much more political and industrial attention inside China in recent years. The prioritization of sufficiency over efficiency, a vestige from China's pre-reform communist ideology of planning economy and a consequence from the slow process of energy marketization and privatization, is still deeply rooted in China's energy apparatus, and will continue to dominate policy-making for a long time.

REFERENCES

China Daily (2011a), 'Industry faces rising power cost', 31 May 2011, accessed 6 November 2017 at http://www.chinadaily.com.cn/china/2011-05/31/content_12608791.htm.

China Daily (2011b), 'Oil reserve project picks up speed', 26 January 2011, accessed 8 January 2016 at http://www.chinadaily.com.cn/business/greenchina/2011-01/26/content_11921393.htm.

China Daily (2012), 'State grid to invest 286 billion yuan in smart grid construction', 25 May 2012, accessed 1 July 2017 at http://www.chinadaily.com.cn/hqcj/zgjj/2012-05-25/content_6005747.html.

China Electricity Council (2011), 'Statistical brief on China's electricity industry in 2010' [*dianli hangye 2010nian fazhan qingkuang zongshu*], accessed 2 March 2018 at http://www.cec.org.cn/huanbao/niandufazhanbaogao/2011-06-27/58873.html.

China's State Electricity Regulatory Commission (2011), *Power Industry Monitoring and Supervision Report 2010* [*dianli jianguan niandu baogao*], Beijing: State Electricity Regulatory Commission.

Cornot-Gandolphe, S. (2014), *China's Coal Market: Can Beijing Tame 'King Coal'?*, Oxford Institute for Energy Studies report, accessed 20 April 2018 at https://www.oxfordenergy.org/wpcms/wp-content/uploads/2014/12/CL-1.pdf.

National Bureau of Statistics of China (2016), 'Statistical Communiqué of the People's Republic of China on the 2015 National Economic and Social Development', 29 February 2016, accessed 20 January 2017 at http://www.stats.gov.cn/english/PressRelease/201602/t20160229_1324019.html.

National Development and Reform Commission of China (2012), *China's 12th Five-Year Plan for Shale Gas* [*yeyanqi fazhan guihua 2011–2015*], accessed 20 April 2018 at http://zfxxgk.nea.gov.cn/auto86/201203/P020120316383507834234.pdf.

National Development and Reform Commission of China (2013), 'NDRC's circular on the promotion of solar PV sector through price leverage' [*guanyu fahui jiage ganggan zuoyong cujin guangfu chanye jiankang fazhan de tongzhi*], 26 August 2013, accessed 8 November 2017 at http://www.ndrc.gov.cn/zwfwzx/zfdj/jggg/201308/t20130830_556127.html.

National Development and Reform Commission of China (2016), *China's 13th Five-Year Plan for Renewable Energy* [*kezaisheng nengyuan fazhan shisanwu guihua*], accessed 2 March 2018 at http://www.ndrc.gov.cn/zcfb/zcfbtz/201612/W020161216659579206185.pdf.

National Energy Administration of China (2012), *China's 12th Five-Year Plan for Coal Industry Development* [*meitan gongye fazhan shierwu guihua*], accessed 2 April 2017 at https://finance.qq.com/a/20120322/005082.htm.

PRC Amended Law on Conserving Energy (2007), accessed 2 December 2017 at https://climateobserver.org/wp-content/uploads/2014/10/Energy-Conservation-Law_2007.pdf.

PRC Law on Conserving Energy (1997), accessed 1 December 2017 at http://www.npc.gov.cn/englishnpc/Law/2007-12/11/content_1383579.htm.

State Council of China (2013), 'State Council circular on China's 12th Five-Year Plan on energy development' [*guowuyuan guanyu yinfa nengyuan fazhan shierwu guihua de tongzhi*], accessed on 1 May 2016 at http://www.gov.cn/zwgk/2013-01/23/content_2318554.htm.

State Information Center of China (2016), 'Targets and approaches of controlling carbon emissions in the 13th Five-Year Plan period' [*shisanwu shiqi tanpaifang kongzhi mubiao ji shixian lujing*], accessed 8 April 2018 at http://www.sic.gov.cn/News/466/6790.htm.

State Information Center of China (2017), 'China facing critical turning point for energy development' [*zhongguo yinglai nengyuanfazhan de zhongyao zhuanxing jiasu qi*], accessed 2 March 2018 at http://www.sic.gov.cn/News/466/8286.htm.

World Nuclear Association (2018), 'Nuclear power in China' (updated March 2018), accessed 2 April 2018 at http://www.world-nuclear.org/information-library/country-profiles/countries-a-f/china-nuclear-power.aspx.

Zheng, L. and L. Mao (2009), 'Nuclear power goal revised to 86GW', *China Daily*, 2 July 2009, accessed 8 December 2017 at www.chinadaily.com.cn/bizchina/2009-07/02/content_8345808.htm.

8. Discussion: energy policy priorities in a fragmented authoritarian state

Chinese President Xi Jinping pledged to build a moderately prosperous society by 2020, with ecological protection being a critical task to achieving this goal. At the closing ceremony of the National People's Congress (NPC) annual session in 2018, Xi vowed to build a beautiful China where the skies are bluer, the land is greener, and the waters are clearer. To change the image of prioritizing development agencies while marginalizing environmental departments in domestic politics, the top leadership in March 2018 restructured the Ministry of Environmental Protection into a new Ministry of Ecological Environment, which took power from the National Development and Reform Commission (NDRC) on issues such as climate change and greenhouse gas emissions.

Nevertheless, it was difficult for the Chinese government to shed overnight its long-lasting development-centric ideology, which has been embedded in its policy-making process and is still underpinned by various interest groups from central to local. China's renewable energy strategy actually is a natural extension of the country's preferred 'no-regret' strategy that emphasizes mitigation actions providing fringe benefits like economic growth and employment to the country, regardless of whether the environmental threat of climate change is real. On the surface, China's coal-dominant energy structure puts it in an awkward dichotomy between energy supply and environmental protection, but in reality, the authoritarian state's obsession with high economic growth, powered by relentless domestic energy production, is the root cause of an economic model characterized by inefficient energy utilization and high-polluting emissions. Such a development-centric ideology of the government and lack of environmental activism at the civic level is preventing China from making full use of the renewable energy facilities installed across the country.

CLASHES BETWEEN PRO-GROWTH AND PRO-ENVIRONMENT INSTITUTIONS

As part of China's energy development policy and climate change/environmental protection policy, the renewable energy policy reflects the country's persistent dilemma in handling the clashes between economic development and environmental protection, both ideologically and institutionally. China's climate change policy per se is part of its energy and environmental strategy, which, in turn, is driven by the country's overall economic growth targets. Climate change 'is an issue involving both environment and development, but it is ultimately an issue of development' China's National Program on Climate Change, Beijing's first comprehensive policy document on the issue, states in its first paragraph (National Development and Reform Commission of China 2007, p. 1). Such a 'development-first' ideology was closely related to the interactive structure of China's inter-agency mechanism on energy and climate change, in which the NDRC and its National Energy Administration (NEA), a macroeconomic and energy management agency, was playing a pivotal role, while the Ministry of Environmental Protection, successor of the State Environmental Protection Administration (SEPA), was marginalized for a long time. Given the profound impact on a broad set of interest groups in the party-state, climate-related policies are often the outcomes of bureaucratic politics and inter-agency competitions within the regime rather than reasoned debates. Inside the Chinese government, with its close focus on the strategic priority of modernizing the economy, economic and energy departments have played a dominant role in shaping the country's climate policy, while environmental administration from the central to local level was in an inferior position in the policy process.

In the 1980s, China treated climate change as a scientific issue and the China Meteorological Administration (CMA) was made responsible for advising the government on policy options in the international arena involving the United Nations Framework Convention on Climate Change. As political awareness and economic sensitivity surrounding climate change increased in the late 1990s, this role passed on to the more powerful State Development and Planning Commission, which has since evolved into the NDRC (Lewis 2007, p. 158). In 1998, China established the National Coordination Committee on Climate Change, an ad hoc inter-agency organization that later included representatives from the NDRC, Foreign Ministry, Ministry of Science and Technology, CMA, SEPA, and ten other ministerial-level departments. With its director appointed as the Chairman of the National Coordination Committee on Climate Change, the NDRC had been managing all routine affairs of the committee and it clearly had a pivotal role in the decision-making process. SEPA ranked fifth in the pecking order within the National Coordination Committee on Climate Change, behind the NDRC, Foreign Ministry, Ministry of Science and Technology, and CMA. Compared with SEPA and CMA, which were potentially sympathetic to the more proactive approach of reducing emissions, the NDRC leans toward rapid economic growth as well as materialized benefits including fund and technology transfers. The main function of the NDRC, a macro-economic regulatory department, is to maximize economic development driven by an expanding energy supply. From its inception, the NDRC was skeptical toward scientific opinion on the causes and impacts of climate change, and it opposed any commitments not in line with economic growth, for example greenhouse gas emission reductions (Richerzhagen and Scholz 2007, p. 318).

Inter-agency conflicts between the NDRC and SEPA subsequently became pronounced as domestic environmental issues increased. An evident inconsistency between the pro-energy commission and pro-environment administration was China's car fuel policy, which went awry in 2007. Vigorously pushed by SEPA, the central government had planned, on 1 July 2007, to introduce a standard for cars known as Euro III, which would reduce carbon monoxide and other emissions. The NDRC, however, in 2007, proposed that the rules should be delayed for up to two years because of the lack of compatible fuel – given that the changes could damage new engines (Chen 2009, p. 22).

In the run up to China's annual parliamentary session of the NPC, *Under the Dome*, an investigative documentary on China's choking air pollution by former China Central Television (CCTV) reporter Chai Jing, had garnered more than 100 million views in less than 48 hours. The documentary, which went viral, interviewed many pro-environment officials, especially those with the Ministry of Environmental Protection and who talked about their inability to regulate polluting companies (*The New York Times* 2015, p. A6), revealing the deep-rooted conflict in policy preferences between economic and environmental bureaucracies. The 104-minute documentary, however, was subsequently censored by the Chinese propaganda department in part due to the fury vented by state-owned energy enterprises. One senior oil company official, Wan Zhanxiang, wrote an essay for Cubeoil.com that attacked Ms Chai's arguments. 'Maybe she doesn't have enough brains and not enough knowledge or thoughts,' he wrote. 'Anyway, she has no insights' (*The New York Times* 2015, p. A6). Thanks to this popular documentary that won applause from Chen but was removed subsequently from websites by China's propaganda department, the environmental issue became one of the hottest topics during the NPC session in 2015. The newly promoted Minister of Environmental Protection Chen Jining compared the video to Rachel Carson's *Silent Spring* in an inaugural news conference on the sidelines of the NPC session.

The high-powered role of the Ministry of Foreign Affairs, Ministry of Commerce, Ministry of Industry and Information Technology, Ministry of Science and Technology, and Ministry of Finance enhances NDRC's advantageous position in the energy and climate change decision-making process (see Figure 8.1). The Ministry of Foreign Affairs puts an emphasis on preservation of national sovereignty, meaning that developing countries have the need and the right to develop and China should not succumb to Western pressures for further mitigation actions. Nevertheless, the Foreign Ministry's concerns about China's national image on the international stage also require the Chinese diplomats to support certain emission-cutting efforts and targets in the domestic political process. The Ministry of Commerce, guided by China's embedded mercantile strategy that seeks trade surplus and foreign exchange reserve accumulation, supports China's participation in the Kyoto Protocol's Clean Development Mechanism (CDM), through which China can benefit from selling emission reduction quotas to other industrialized nations. The Ministry of Commerce is also keen on supporting Chinese manufacturers of wind turbines and solar panels that can be shipped to international markets. It is not enthusiastic, however, about the stimulation of domestic demand for such renewable energy products. The Ministry of Industry and Information Technology and Ministry of Science and Technology' focus on research and development of indigenous low-carbon technology and foreign technological aid and transfer in this regard.

Also heavily influenced by the mercantile principle, the Ministry of Finance is in charge of giving grants and subsidies to all the sectors related to indigenous technological research, manufacturing of parts and equipment, and power generation in the fossil fuel and low-carbon areas. Without any structural change to the inter-agency energy and climate change mechanism, the 'development-first' views of these pro-growth bureaucracies will undoubtedly continue to shape the major features of China's renewable energy policy. Under pressure from the international community on account of its rapidly growing carbon emissions, China decided to upgrade the existing inter-agency climate mechanism to a higher level in 2007, when the National Leading Group on Energy Saving, Emission Cutting and Climate Change was set up and headed by the then Premier Wen Jiabao himself instead of a ministerial-level official from the NDRC. The National Leading Group on Energy Saving, Emission Cutting and Climate Change put more emphasis on the environment aspect in response to the grim climate context, but the NDRC's dominant position in the policy-making process has not changed significantly, as the National Leading Group still housed its administrative office within the NDRC.

The pro-growth feature concerning China's climate policy-making mechanism remained untouched due to the lack of significant structural changes inside the National Leading Group, which was headed by the Premier, whose principal task was to boost the economy and create more jobs, and his two associates, a Vice Premier with an NDRC background and a State Councilor promoted from the post of Foreign Minister. The NDRC chief was still the director of the National Leading Group's administrative office in charge of coordination work among various ministries and commissions. International climate change politics after 2009 became a finger-pointing game, in which China often found itself being blamed for irresponsibility or leading accusations that rich nations were trying to sabotage climate talks. The grim fact that China had become the world's largest carbon emitter seemed to have had a substantial impact on the global public's psychology in an era characterized by growing climate consciousness. Given a bleak international climate outlook and China's alarming energy and emission trajectory, Chinese leaders kept open the possibility of making larger concessions to ensure that Chinese emissions peaked and then declined within two decades. In China's domestic politics, governmental restructure was carried out to terminate the NDRC's dominance over climate change-related policies in March 2018, when its decade-long bureaucratic rival, the Ministry of Environmental Protection, was revamped into the Ministry of Ecological Environment, a new cabinet ministry in charge of issues related to climate change and greenhouse gas emissions.

Although the Ministry of Ecological Environment has replaced the NDRC as the pivotal administration in China's climate change policy-making, the NDRC and its semi-autonomous NEA remain powerful in making energy-related policies and plans, including those regarding non-fossil fuels. Despite Xi Jinping's 'new normal' slogan that tries to strike a balance between growth and environment, China's maintenance of more-than-6-percent economic growth in his tenure continues to impose huge pressure on natural resources and the ecology. In compensation for the NDRC's handover of the climate change-related power to the Ministry of Ecological Environment, NDRC Director He Lifeng in March 2018 was awarded a state-leader position in the Chinese People's Political Consultative Conference (CPPCC), which granted him special political privileges vis-à-vis other cabinet ministers, including the Minister of Ecological Environment.

COMPETITION BETWEEN THERMAL AND LOW-CARBON POWER SECTORS AND THE ROLE OF GRID

In the process of energy policy-making, related interest groups exert their influences on the central government through their formal and informal connections to various government departments, with the NDRC/NEA being the top lobbying target and other economic ministries also under pressure from other institutional players (Figure 8.1). In most cases, the bureaucracy at the central and local levels cannot ignore the voices of the interest groups with strong government backgrounds, including large state-owned enterprises (SOEs), such as the three national oil companies (China National Petroleum Corporation (CNPC), China Petrochemical Corporation (Sinopec) and China National Offshore Oil Corporation (CNOOC)), state-owned coal mining companies such as the Shenhua Group and China National Coal Group, the State Grid Corporation, and state-owned power generation companies including the Big Five generators (mostly thermal), China Three Gorges Corporation (hydropower) and China National Nuclear Corporation (nuclear) (see Chapter 5).

Among all the institutional pressures on energy policy formulation, those from fossil fuel and electricity-transfer business juggernauts are highly influential due to their traditional and existing ties with the NDRC and the NEA. The three NOCs were the outcome of restructuring of the Ministry of Petroleum and Ministry of Chemical Industry in the 1980s, while the China Electricity Regulatory Commission that oversaw the power industry as well as the Big Five generators were spin-offs of the defunct ministerial-level State Power Corporation. In 2013, the China Electricity Regulatory Commission was absorbed into the NEA. Most of the wind power or solar power generation plants are part of the state-owned power generation giants, which still mainly rely on thermal power and are not enthusiastic about lobbying government officials through their connections with the bureaucracy to prioritize renewables.

Conventional energy giants, including the fossil fuel producers, thermal power generation companies, hydropower corporations and nuclear power plants, mainly exert their influences on China's energy policy-making through lobbying activities targeting the NDRC and the NEA under the NDRC. Wind and solar power plants not only need to lobby the NDRC and NEA for favorable energy development plans, but also are sensitive to supportive policies from the Ministry of Finance and Ministry of Industry and Information Technology (see Figure 8.1). Companies producing wind turbines, and solar modules and cells are lobbyists at the Ministry of Finance for grants and subsidies, the Ministry of Industry and Information Technology for long-term industrial policy support, the

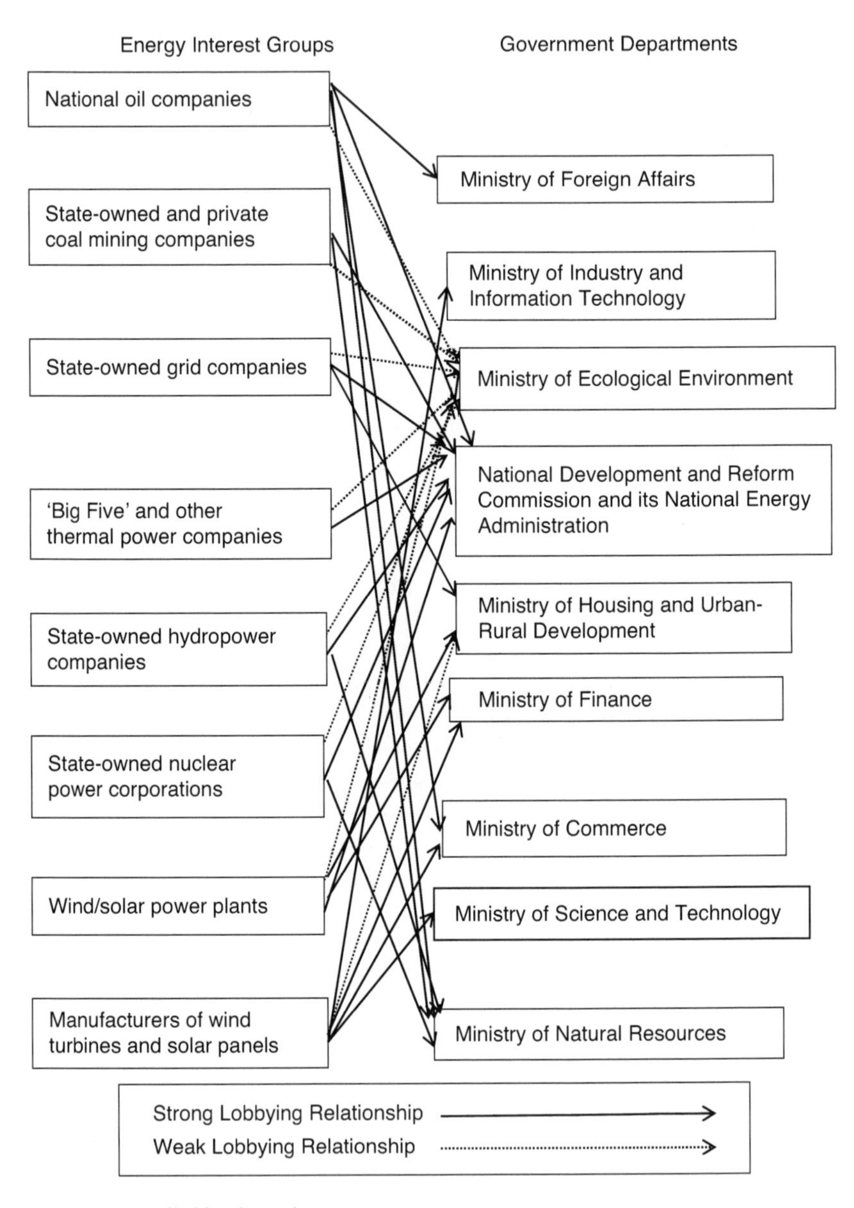

Source: compiled by the author.

Figure 8.1 Energy interest groups and their lobbying targets at the central level

Ministry of Science and Technology for research and development (R&D) support, and the Ministry of Commerce for export subsidies and trade dispute settlements. With regards to the implementation of renewable energy policies at local levels, it is subject to the interests of local governments and the level of coordination between local governments and central government. The central government is responsible for policy regulation while the local governments are in charge of policy implementation.

Unlike in the past when local levels used to be subject to a high degree of control from the central government, after the economic reform and transition from a planned economy to a market economy, such control has been diminishing and local government have become more politically and economically independent. As they have their own interests to pursue, they need tactically to strike a balance between developing the local economy and serving as political agents of the central government. Sometimes, they prioritize local economic development and chose non-compliance or partial compliance with the orders from central government. Sometimes the coordination between central government and local governments has also been poor and it limits the efficiency of policy implementation at local level (Zhao 2001, p. 6).

As local governments have different priorities, sometimes they may justifiably reinterpret national policies and combine them with more pressing issues that are of greater importance to local development; sometimes they may actively or passively obstruct the implementation of central government when the policy is not in their development interest; or sometimes it is simply because the local government lacks the resources to implement national policies (Meidan et al. 2009, p. 616). In most cases, local governments are more interested in the installation of wind and solar power generation facilities instead of getting them to operate in full swing. The related infrastructure building, land allocation, and equipment procurement not only contribute to the growth of the local gross domestic product (GDP) and provide opportunities for rent-seeking and payoffs for local officials, but also increase employment and serve the social function of maintaining stability. Far more workers are employed in areas that underpin the use of solar technology, such as making steel racks that angle the panels toward the sun. The bulk of workers in the solar industry install and maintain the projects, a process that is labor-intensive and hard to automate (*The New York Times* 2018, p. 8).

China is a global leader in constructing ultra-high-voltage (UHV) transmission lines, a technology that can carry electricity across vast distances with great efficiency and thus help alleviate power generation overcapacity in its interior provinces. Nevertheless, China's enthusiasm for UHV has been waning recently due to conflicts of interest between grid companies and the central and local governments (Chinadialogue 2018b). Both the central government and two major state-owned grid companies, State Grid and China Southern Power Grid, are interested in building UHV networks connecting the various provinces, but their policy priorities are different – the central government planners want the UHV to carry more electricity from non-hydro renewable sources such as wind and solar, while the grid companies, driven by commercial interest embedded in the long-established coal and hydropower industries, prefer using UHV to transfer electricity from thermal or hydropower plants.

In 2016, UHV lines transported 248.78 TWh of electricity, 74 percent of which was from non-fossil fuel sources, mainly hydropower (Table 8.1).

Table 8.1 Electricity transmitted by China's UHV lines in 2016

UHV Line	Sources of Power	Total Power Transmitted (TWh)	Power from Renewable Sources (including hydro) (TWh)	% of Renewable Power (including hydro) in Total
Changnan Line (State Grid)	thermal & hydro	8.25	2.92	35%
Ximeng Shandong (State Grid)	thermal	3.28	0	0
Wandian Dongsong (State Grid)	thermal	20.29	0	0
Zhefu Line (State Grid)	thermal	17.1	0	0
Fufeng DC (State Grid)	hydro	32.61	32.48	100%
Jinsu DC (State Grid)	hydro	38.33	38.25	99%
Binjin DC (State Grid)	hydro	36.75	36.75	100%
Tianzhong DC (State Grid)	thermal, wind and solar	32.26	7.34	23%
Lingshao DC (State Grid)	thermal, wind and solar	7.28	2.08	29%
Chusui DC (China Southern Power Grid)	hydro	26.18	26.18	100%
Puqiao DC (China Southern Power Grid)	hydro	26.45	26.45	100%
Total		248.78	172.45	74%

Source: Reports from the State Grid and China Southern Power Grid (BJX.com.cn, 2017).

Only 9.42 TWh of non-hydro renewable electricity was transferred by two UHV lines, namely Tianzhong DC line and Lingshao DC line (Table 8.1), accounting for 4% of total power transferred by UHV lines in China (BJX. com.cn 2017). Moreover, most of the newly-constructed UHV lines are far from reaching their full transmission capacity, with the Zhefu line and Ximeng-Shandong line, both under the management of the State Grid, having transmitted 1.71 TWh and 3.28 TWh respectively in 2016, well below their targeted goal of transmitting 40 TWh per year (Chinapower. com.cn 2017).

Some provincial governments targeted by the central government for UHV projects have been skeptical, thanks to their own concerns about local employment and fiscal revenues brought about by local power plants. Local authorities are often more interested in building their own power plants rather than importing power from other provinces through UHV lines. The construction of the UHVDC Sichuan Number Four line was included in China's 13th Five-Year Plan (2016–20) to take hydropower generated in the southwestern Sichuan province to the eastern Jiangxi province. Nevertheless, such a central plan was opposed by Jiangxi province because Jiangxi officials wanted locally generated power instead. In practice, Jiangxi province has been trying to support the building of new coal-fired power plants connected to the local grid while delaying the operation of the new UHV line until after 2025 (Chinadialogue 2018b). The central province of Hubei, one of China's traditional industrial bases, has also been reluctant to accept the long-debated UHV lines that connect it with energy-rich northwestern or southwestern China.

China is undergoing an enormous energy transition as the country moves away from coal, which at the end of 2016 still comprised 62 percent of the energy mix (China's National Bureau of Statistics 2017), and towards renewable and clean sources of energy. However, such a transition has met strong resistance from the fossil fuel sectors, especially the coal-related sectors. In the context of China's growing energy consumption, its coal consumption had fallen three years in a row by 2016. In 2016 alone, China's coal consumption dropped 4.7 percent year on year, while the proportion of 'clean energy' (hydropower, nuclear, wind, solar, natural gas) in the energy mix rose to 19.7 percent, up 1.7 percentage points from a year ago (China's National Bureau of Statistics 2017). Government departments and grid dispatchers with vested coal interests have been accused of exerting administrative power to disrupt wind production. When designing renewable energy policies, the Chinese policy-makers expect local governments to play a cooperative role in enforcing those policies, laws, and regulations. In practice, however, this high degree of administrative cohesion does not exist in places where local government interests diverge sharply from those of the central planners, and they create substantial obstacles to strict enforcement of both national and local energy policies.

LOBBYING FOR FAVORABLE ON-GRID TARIFFS AND OTHER SUPPORTIVE POLICIES

China's power industry, despite being increasingly competitive and fragmented as a result of government-initiated structural reforms (Chapter 6), is still highly regulated by the state, with the on-grid electricity tariff, in most cases still set by governmental regulators, being vital to profit margins and even survival of various power generators. The 'on-grid electricity tariff', therefore, has been the main focus of the lobbying activities of a variety of interest groups. Compared with the major hydropower and nuclear power corporations, which are state-owned giants subject to state plans and government edicts, the wind and solar power plants, which are mostly small- and medium-sized generators with diverse ownership, are more sensitive to fluctuations in power pricing, production costs, and demand conditions.

Lobbyists are also interested in influencing government energy plans formulated by the NDRC. These plans, which are more precise in predicting future development trajectories of conventional energy, often lag behind the unexpected proliferation rates of wind and solar power plants that face higher supply and demand elasticity under conditions of government subsidies, feed-in tariffs, and a slump in production costs. To foster the development of renewable energy resources, differentiated on-grid electricity tariffs have been adopted to provide incentives for renewable power producers including wind and solar.

Besides feed-in tariffs, China's Renewable Energy Law has also introduced other schemes such as the 'renewable portfolio standard' (RPS) and 'mandatory grid-connection' that had been successful in advancing the cause of renewable energy in Europe and North America. It is evident from the Chinese version of RPS – which had originally included plans to increase the proportion of renewable energy (including large hydropower) in primary energy consumption from 7.5 percent in 2005 to 15 percent in 2020 – that the Chinese government has prioritized the development of wind power over solar power generation in the mid-term. The NDRC allocated the aforementioned RPS target (excluding hydropower) as 30 GW from wind, 30 GW from biomass, and 1.8 GW from solar photovoltaics (PV) (Zhang 2005, p. 1), of which solar PV would be projected to produce only 6 percent of the total wind power generation. The Chinese government subsequently revised the RPS goal upwards, with the intention to produce 20 percent of the country's energy requirements from renewable sources by 2020, increasing the target of wind power capacity to 100 GW and retaining the solar capacity target at 1.8 GW.

The Renewable Energy Law's 'mandatory grid connection' clause required grid companies to 'buy the grid-connected power produced with renewable energy within the coverage of their power grid' and to 'provide grid-connection service for the generation of power with renewable energy' (PRC Renewable Energy Law 2005, Article 14). To help grid companies share such costs with end-users, the law allowed grid companies to recover the grid connection expenses related to the purchase of renewable power and other reasonable expenses in power transmission costs from the selling price (PRC Renewable Energy Law 2005, Article 21). After four years of implementing the Renewable Energy Law, despite China's efforts to rapidly increase its installed capacity of renewable energy, the government realized that much of this capacity was not promptly connected to the grid and that not all power being generated was being purchased as required by the law.

When the Renewable Energy Law was amended in 2009, the top legislature required electricity grid companies to buy all the power produced by renewable energy generators, and authorized the State Council energy department – in conjunction with the state power regulatory agency and the State Council finance departments – to determine the percentage share of electricity generated from renewable energies in the total quantity of electricity generated during the planned period (PRC Amended Law on Renewable Energy 2009, Article 14), a regulation system similar to the renewable portfolio standards adopted by some states in the United States. The amendment stipulated that power enterprises refusing to buy power produced by renewable energy generators would be fined up to an amount double that of the economic loss of the renewable energy company (PRC Amended Law on Renewable Energy 2009, Article 29). To encourage grid companies to accept and purchase more power generated from renewable sources, the amendment partially waived the value-added and income taxes levied on grid companies on their revenue (the tax rate is about one-third of the revenue) generated from the surcharge on the retail power tariff for supporting renewable energy.

STATE VERSUS SOCIETY

State–society relations are defined as interactions between state institutions and societal groups to negotiate how public authority is exercised and how it can be influenced by people (Department for International Development 2010, p. 15). When the Chinese government started to extend its reform and opening up in the 1980s, civil activities became frequent and all kinds of non-governmental organization (NGO) working for public and special interests mushroomed. Although many of these NGOs have also become increasingly visible players in China's energy- and environment-related politics, their roles and functions are still quite restrained under China's one-party political system. Therefore, they are not as strong and influential as their international peers. In response to the growing civil demand for greater participation in ecological protection, the Chinese government has robustly supported the dominant roles played by government-organized NGOs (GONGOs) in promoting environmentally friendly activities and renewable energies. In China, GONGOs definitely enjoy much more political and financial advantages and policy convenience compared to other NGOs, but this kind of government backing also hinders their role as a third force against the state and market.

In China's state-centric society, the influence of environmental NGOs (ENGOs) is still very weak and the role of the media is also quite limited, because the government imposes strict regulations on them. Based on the experiences in industrialized nations, the voice of these social organizations and civic society play a vital role in identifying the key design and implementation features that maximize synergies between environmental protection and economic development. They are focused on issues such as defining the mutual rights and obligations of state and society, negotiating how public resources should be allocated, and establishing different modes of representation and accountability (Department for International Development 2010, p. 15). The focus is not on particular institutional forms but rather on the relations and relational functions of state and societal institutions. Neither the state nor civil society is seen as acting in isolation. Rather, the state derives its legitimacy through its interaction with citizens and an organized and active civil society. The lack of effective civic supervision and lobbying activities impedes the enforcement of China's low-carbon energy ambition and other environmental initiatives, as political and business elites, who benefit from patronage and income from natural resources, rents, and favorable policies, often have little incentive to engage with citizens and to build effective public authority over energy and environmental issues. The concentration of power in a few elites at both central and local levels also limits the participation of ENGOs and citizens in promoting environmentally friendly production or consumption. In recent years, although ENGOs and the emerging civil society have become increasingly visible players in China's green politics, their roles and functions are still marginal under China's authoritarian political system.

Friends of Nature, China's first ENGO, was established in 1994, when only nine ENGOs were founded in China, including four student organizations. By 1996, the number had grown to 28, including ten student organizations. The number increased dramatically between 1997 and 1999, when at least 69 ENGOs were founded, 43 of which were student organizations (Yang 2005, p. 50). In 1998, in response to the sharp increase in the number of NGOs in China, the government promulgated regulations on the registration and management of NGOs and non-profit organizations, which required any NGO based in China to be registered at the Ministry of Civil Affairs and to have a sponsoring governmental institution (Jin 2001, pp. 5–8). The number of ENGOs grew 38.8 percent annually between 2007 and 2012 in China, reaching 7,881 by the end of 2012, 6,816 of which were registered as social organizations and 1,065 as non-profit organizations (*People's Daily* 2013). Before 2008, the main target of activities to prevent environmental pollution by grassroots NGOs had been individual polluting corporations, but later, grassroots ENGOs began to adopt more flexible approaches, with corporations being their sponsors from the beginning (Aikawa 2017, p. 190).

China's officially registered NGOs are categorized by the Ministry of Civil Affairs into three groups: social organizations (*shetuan*), private non-enterprise units (*minban feiqiye*), and foundations (*jijinhui*). When these NGOs are being scrutinized in terms of their backgrounds and features, seven categories of social groups can be generally identified: GONGOs, other register NGOs and foundations, non-profit enterprises that are registered as business entities but operate as non-profit organizations, unregistered voluntary groups, web-based groups, student associations that are registered with campus youth leagues but function as NGOs, and university research centers (Yang 2005, p. 50). Chinese ENGOs can also be divided into two classes based on their origins: international and domestic (Aikawa 2017, p. 178). In this state-centered country, GONGOs like the All-China Environment Federation, the China Association of the Environmental Protection Industry, the China Society of Environmental Sciences, the China Energy Conservation Association, and the Chinese Society for Sustainable Development definitely enjoy much more financial advantages and policy convenience compared with other NGOs, but as noted before this kind of government support also limits their roles as a third force against the state and market.

International ENGOs came to China with a common aim to promote environmental consciousness and public participation, avoiding confrontational methods and adopting approaches that encourage learning, cooperation, and participation. Greenpeace East Asia has for years been in collaboration with some Chinese industry associations and research groups with regard to promotion of renewable energy, calculating the potential co-benefits of wind and solar PV power generation in China by releasing the report *Accelerating the Energy Transition: The Co-Benefits of Wind and Solar PV Power* in 2017. The findings of the report were evidence of the huge social, economic, and environmental benefits that wind and solar PV can bring to China. The findings countered arguments that renewables will be costly and bring undue burden to the electricity grid, as well as fears that the energy transition will result in unavoidable social and economic losses due to coal mine and coal power plant closures (Greenpeace 2017).

Another influential player in promoting environmental consciousness and low-carbon solutions is Chinadialogue, an independent organization dedicated to promoting a common understanding of China's urgent environmental challenges as well as a non-profit and bilingual (Chinese and English) website launched in 2006 to warn against risks of climate change, species loss, industrial pollution, water scarcity, and environment damage pertaining to China (Chinadialogue 2018a). Chinadialogue has paying consistent attention to China's energy and climate change issues, including the tension between the powerful coal industry and renewable alternatives, high level of 'curtailment' in wind power plants (Chinadialogue 2016) and the disappointing impact of UHV power lines in transferring the redundant wind and solar power from northwestern China (Chinadialogue 2018b).

Local ENGOs in China pay more attention towards prevention of industrial pollution rather than low-carbon and energy saving solutions. Besides, it is still very difficult for ENGOs to attract funding as well as talented professionals under the current policy environment, in which local officials still suspect the real motives of NGOs and volunteers and are often hostile towards these social forces. Compared with GONGOs, which have been set up as the regime's backbone social organizations and are thereby privileged to participate in state-centric environment and mitigation effort, grassroots NGOs still have a large untapped potential to become efficient pressure groups in local energy or climate-related politics. Moreover, the independence of grassroots NGOs has been endangered with the tightening of regulations. In the latter half of 2014, the Communist Party of China (CPC) began to require every NGO to set up a CPC branch within the organization, which means that every ENGO must allow one or more Party members to join the organization (Aikawa 2017, pp. 193–4).

Local Chinese ENGOs, whether registered or unregistered, web-based or driven by student activism, all have their own weaknesses in capacity-building and internal management. Although most Chinese NGOs have become more mature and sophisticated over the years, they still face insufficiency and weakness in their managerial structure and professional capacity, which prohibits them from playing a larger role in curbing pollution and promoting sustainability. In many of China's first generation of grassroots ENGOs, management and function evolved spontaneously, with the organizations being too dependent on their founder-leader. The leaders' personal ambitions and charisma shaped these organizations and mobilized followers. Management was not institutionalized and decision-making remained concentrated in the hands of a small group of individuals (Fu 2007, p. 307). These NGOs need to institutionalize their managerial structure and improve internal evaluation systems to promote administrative efficiency.

As today's energy- and climate-related issues become more and more complex, the solutions to these problems demand expertise in science, technology, economics, finance, and other areas. Lack of professional knowledge in these areas will prevent these organizations from playing an active and effective role in influencing government decision-making and the setting of public agendas. More often than not, grassroots ENGOs, when facing pressures from local- or national-level interests that favor consumption of polluting energy sources or abandonment of clean energy, do not have capacity to cope with challenges either from state sectors or organizations having close ties to the state apparatus. Under such circumstances, NGOs need to take a professional rather than a general perspective to effectively impact local or national energy/climate issues. The time of relying on a certain charismatic leader has passed and there is an urgent need for Chinese NGOs to standardize their operational and managerial procedures and lift their professional standards.

In the absence of a favorable policy situation in China, however, it is difficult for NGOs to attract funding as well as talented professionals. China's restrictive regulatory system concerning registration, organization, and funding is still preventing most of these NGOs from carrying out substantial supervisory missions. The Ministry of Civil Affairs prohibits most NGOs from establishing subsidiary organizations in other areas, and from engaging in public fund-raising independently. The de facto restrictions from the government, therefore, impose an indirect influence on capacity-building of ENGOs in China. Despite the large number of NGOs in the country, they can only play very limited roles in supervising the government and setting agendas in the development of renewable energy.

In recent years, the Chinese government further tightened its control over NGOs, social media, and other civic activism. Besides the propaganda department's censorship of Chai Jing's *Under the Dome* documentary that revealed conflict between energy and environmental sectors in 2015, the central authorities also delegitimized a number of ENGOs in 2016, including the China Renewable Energy Industry Association, China Clean Energy Industry Association, China New Energy Association, and China Energy and Environmental Industry Association (Ministry of Civil Affairs of China 2016). The crackdown was part of the central government's effort to disband 'copycat and offshore social organizations' (*shanzhai he li'an shetuan*), which were identified by the Ministry of Civil Affairs as 'carrying out illegal activities and violating rules' (*China Daily* 2016, n.p.).

The China Renewable Energy Industry Association, among others, had engaged in publishing magazines and organizing public forums on renewable energy development, sometimes with endorsement from the NEA, the NDRC, or the Ministry of Commerce. According to the Ministry of Civil Affairs, the closedown of some of these social organizations was based on requests from local public security departments and the industrial and commercial sectors (*China Daily* 2016), which indicate the vulnerability of these social organizations when confronting powerful interest groups, even if they occasionally also have links with governmental departments. Chinese NGOs and social media do not necessarily represent the voices of civic society on all occasions. They can be manipulated by different interest groups within or outside the government to maneuver online opinions, consolidate their own power and interest, undermine their rivals' reputation and credibility, and set energy or environmental agendas in favor of their interests.

The Chinese government on the one hand has enhanced regulations on rampant activities on the Internet and social media public relations (*wangluo tuishou*), a type of organized business operation in which online posters are recruited to use multiple Internet identities in promotional and marketing campaigns, particularly in Internet forums and social networking sites (Wu 2011, pp. 1–2). On the other hand, the state actors, based on lessons learnt from previous setbacks in the early development of social media, has been making efforts to maneuver online opinions in public relations campaigns initiated by their own propaganda teams. As for the renewable energy issue, more and more Chinese companies are showing an interest in corporate social responsibility (CSR), which concerns the relationship not just with their clients, suppliers, and employees, but also with other social groups, and with the needs, values, and goals of the society in which they operate (United Nations 2000, p. 2; Twigg 2001, p. 5).

Like their peers in developed economies, business managers in China have committed themselves to CSR that covers a wide range of public interest areas like ecological protection, renewable energy development, and emissions reduction. As carbon and other pollutant emissions pose a serious threat for local sustainable development and cause substantial environmental damage, the low-carbon solutions to the challenges to a large extent lie in CSR that could be provided by energy corporations in parallel with the state and civil society. Pro-wind and solar sectors have been focusing on the carbon intensity of fossil fuels in the climate change context, and the ecological hazards hydro and nuclear power plants could bring about. At the same time, conventional energy sectors, via GONGOs or other social organizations such as the China National Coal Association, China Water Enterprises Confederation, and China Nuclear Energy Association, have been justifying their own industries from the CSR perspective and raising concerns over the huge government subsidies given to renewable sectors. In an increasingly pluralistic domestic environment, these NGO and media voices supporting different energy solutions are having non-negligible impacts on China's renewable energy policy formation and implementation.

Compared with other democracies, the authoritarian one-party regime in China should face much less resistance from civil society as it pushes forward those mammoth hydropower and nuclear projects that may pose potentially serious threats to the local ecosystem. But even so, the environmental impact of power plants is a growing concern that affects the government's decision to approve large-scale projects. The construction of the Three Gorges hydropower complex had provoked a long-lasting debate over the pros and cons of this largest-ever hydropower project. In 1992, when the then Chinese Premier Li Peng submitted a proposal for the Three Gorges Dam to the NPC, which is usually seen as pliant, there was strong opposition against the proposal, and this was ultimately passed with the smallest margin in the legislature's history.

China passed the Law on Evaluation of Environmental Effects in 2002, requiring stakeholders of all power projects constructed within its territory to submit a comprehensive evaluation of environmental impacts on the ecosystem before construction begins (PRC Law on Evaluation of Environmental Effects 2002, Articles 3 and 4). The law, which gives state environmental regulators veto power over the approval of all kinds of new projects, including energy projects, encourages stakeholders, specialists, and the public to participate in the evaluation of environmental effects (PRC Law on Evaluation of Environmental Effects 2002, Article 5). With increasing public awareness of the negative environmental effects associated with large-scale projects, especially gigantic hydropower stations, the Law on the Environmental Impact Assessment has been ratified since 2002 and led to a slowdown in hydropower expansion in subsequent state plans. In 2003 and 2004, proposals to build a string of dams on the southwest Nu River (also known as Salween) sparked a fierce debate in

China. Diverse reporting perspectives and heated debates in the mass media therefore shaped public opinion, which in turn influenced government policies. Consequently, in April 2004, the State Council halted the construction of the River Nu hydropower project.

Previously, radiation accidents that happened elsewhere also raised safety concerns about nuclear power, and this explains why China has lagged behind in the nuclear power sector. Such anxiety dissipated in the early 2000s as energy demand soared and China gained confidence in operating its own nuclear power plants. Nevertheless, China suspended nuclear plant construction nationwide for 20 months after the onset of the Fukushima Daiichi nuclear disaster in Japan, in response to public concerns about nuclear safety. Although the government resumed nuclear plant construction along the eastern coastal area in late 2012, most of the planned nuclear projects located in the hinterland, which may be susceptible to greater disastrous consequences such as radioactive leakages, are still suspended. Such environmental concerns raised by societal forces subsequently persuaded central planners to tilt towards wind and solar power generation in terms of supportive policies and growth targets (see Chapter 7) in China's Five-Year Plans.

REFERENCES

Aikawa, Y. (2017), 'Environmental NGOs and environmental pollution in China', in H. Kitagawa (ed.), *Environmental Policy and Governance in China*, Japan: Springer Japan KK, pp. 177–94.

BJX.com.cn (2017), 'How much renewable energy was transmitted by 11 UHV lines?' [*11tiao tegaoya xianlu daodi shusongle duoshao kezaisheng nengyuan?*], accessed 19 February 2018 at http://news.bjx.com.cn/html/20170424/821851.shtml.

Chen, G. (2009), *Politics of China's Environmental Protection: Problems and Progress*, Singapore: World Scientific.

China Daily (2016), 'Ministry exposes copycat and illegal social organizations', 28 April 2016, accessed 1 February 2018 at http://www.chinadaily.com.cn/china/2016-04/28/content_24943739.htm.

Chinadialogue (2016), 'How China can stop wasting wind energy', 22 July 2016, accessed 1 December 2017 at https://www.chinadialogue.net/article/show/single/en/9119-How-China-can-stop-wasting-wind-energy.

Chinadialogue (2018a), 'Our mission', accessed 5 January 2018 at https://www.chinadialogue.net/pages/about.

Chinadialogue (2018b), 'Sparks fly over ultra-high voltage power lines', 29 January 2018, accessed 10 February 2018 at https://www.chinadialogue.net/article/show/single/en/10376-Sparks-fly-over-ultra-high-voltage-power-lines.

Chinapower.com.cn (2017), 'Wind and solar power not popular' [*fengdian guangfu bingbu fengguang*], 25 April 2017, accessed 18 February 2018 at http://www.chinapower.com.cn/informationhyfx/20170425/74216.html.

China's National Bureau of Statistics (2017), *Statistical Bulletin of China's Economic and Social Development for 2016* [*zhonghua renmin gongheguo 2016nian guomin jingji he shehui fazhan tongji gongbao*], accessed 15 February 2018 at http://www.gov.cn/xinwen/2017-02/28/content_5171643.htm.

Department for International Development (DFID) (2010), *Building Peaceful States and Societies: A DFID Practice Paper*, London: Department for International Development, accessed 14 February 2018 at http://www.gsdrc.org/docs/open/con75.pdf.

Fu, T. (2007), 'Development of environmental NGOs in China', in C. Liang and D. Yang (eds), *The China Environment Yearbook 2005: Crisis and Breakthrough of China's Environment*, Boston: Brill, pp. 291–310.

Greenpeace (2017), 'By 2030 China's wind and solar industry could replace fossil energy sources to the tune of 300 million tonnes of standard coal per year', press release, 11 April 2017, accessed 1 May 2018 at http://www.greenpeace.org/eastasia/press/releases/climate-energy/2017/By-2030-Chinas-wind-and-solar-industry-could-replace-fossil-energy-sources-to-the-tune-of-300-million-tonnes-of-standard-coal-per-year/.

Jin, J. (2001), 'The growing importance of public participation in China's environmental movement', in J. Turner and F. Wu (eds), *Green NGO and Environmental Journalist Forum: A Meeting of Environmentalists in Mainland China, Hong Kong, and Taiwan*, Washington, DC: Woodrow Wilson Center, pp. 5–8.

Lewis, J. (2007), 'China's strategic priorities in international climate change negotiations', *The Washington Quarterly*, 31 (1), 155–74.

Meidan, M., P. Andrews-Speed, and M. Xin (2009), 'Shaping China's energy policy: actors and progresses', *Journal of Contemporary China*, 18 (61), 591–616.

Ministry of Civil Affairs of China (2016), 'A list of copycat and offshore social organizations exposed by the Ministry of Civil Affairs', 27 April 2016, accessed 1 January 2018 at http://www.mca.gov.cn/article/zwgk/mzyw/201604/20160400000066.shtml.

National Development and Reform Commission of China (NDRC) (2007), *China's National Climate Change Program*, Beijing: NDRC.

People's Daily (2013), 'Around 8000 environmental NGOs in China', 5 December 2013, accessed 1 December 2017 at http://paper.people.com.cn/rmrbhwb/html/2013-12/05/content_1357312.htm.

PRC Amended Law on Renewable Energy (2009), accessed 5 January 2018 at www.npc.gov.cn/npc/xinwen/2009-12/26/content_1538199.htm.

PRC Law on Evaluation of Environmental Effects (2002), accessed 2 November 2017 at http://www.npc.gov.cn/englishnpc/Law/2007-12/06/content_1382122.htm.

Richerzhagen, C. and I. Scholz (2007), 'China's capacities for mitigating climate change', *World Development*, 36 (2), 308–24.

The New York Times (2015), 'China blocks web access to video about country's rampant air pollution', 7 March 2015, p. A6.

The New York Times (2018), 'Tariffs cloud U.S. solar industry's future', 25 January 2018, p. 8.

Twigg, J. (2001), *Corporate Social Responsibility and Disaster Reduction: A Global Overview*, accessed 4 June 2017 at http://r4d.dfid.gov.uk/PDF/Outputs/Mis_SPC/R7893CSROverview.pdf.

United Nations (2000), *Development of Guidelines on the Role and Social Responsibilities of the Private Sector*, Report of the Secretary General to the Preparatory Committee for the special session of the UN General Assembly on the World Summit for Social Development and Beyond (A/AC.253/21), New York: United Nations.

Wu, M. (2011), *China's Crackdown on Illegal Internet Publicity Activities*, Background Brief No. 681, Singapore: East Asian Institute (EAI).

Yang, G. (2005), 'Environmental NGOs and institutional dynamics in China', *China Quarterly*, 181, 46–66.

Zhang, G. (2005), 'How can energy shortage be blamed on China?', *People's Daily Overseas Edition*, 21 September 2005, p. 1.

Zhao, J. (2001), *Reform of China's Energy Institutions and Policies: Historical Evolution and Current Challenges,* BCSIA Discussion Paper 2001–20, Boston, MA: Energy Technology Innovation Project, Kennedy School of Government, Harvard University.

9. Conclusion: policy priorities reshaped by central–local relationship and interest groups

Through examining China's recent renewable energy development, shaped by the dynamic interactions among various interest groups, central–local relationships, and the central government's energy policy priorities, this book questions authoritarian environmentalists' prevailing view of hailing China as a role model in promoting low-carbon energy. China's prioritization of renewable energy production, originally driven by international climate change pressures and reflecting its mitigation commitment, has, however, been discounted by influences of domestic interest groups and a complicated central–local relationship. The scrutiny of policy priority variation over time also reveals the mercantile features of the central's renewable power strategy, which, in many cases, are given precedence over the country's environmental determinations. Pro-development and pro-environment disputes, at both ideological and bureaucratic levels, have led to such policy inconsistency and confusion.

POLICY INTERACTIONS BETWEEN CENTRAL AND LOCAL GOVERNMENTS, INTERNATIONAL PRESSURE, INTEREST GROUPS, AND SOCIETAL FORCES

This book, with its focus on the redundant capacity in China's power generation sectors and supporting industries, to a large degree helps to explain the institutional and ideological factors that have led to China's widespread overcapacity in most industrial sectors. The discussion in the preceding chapters revealed that China's state capitalism has an innate tendency of focusing on the 'supply side' instead of 'demand side' (Kang and Su 2016, p. 16), which differs from a market-driven economy. This has resulted in enormous industrial capacity subsidized by the state but detached from real market demand. Under Xi Jinping's economic 'new normal' ideology emphasizing growth shifting to a sub-high gear, economic distortions that were concealed during the high growth period have now surfaced to constrain further growth (Li and Quan 2017, p. 139).

The 'supply side' reform advocated by Xi's advisors aimed to reduce redundant industrial capacity through closure of factories and investment controls. In reality, the reform only had limited impact on local capacity expansion since mercantile policy priorities defined by central–local relationships and interest groups remained basically intact. To fundamentally solve the huge waste problem associated with overinvestment and overcapacity, China has to transform its economy from a top-down and state-driven system to a new stage driven by domestic demand and with fewer policy interventions from the governmental apparatus at various levels.

China's nascent development of renewable energy and supporting industries was largely motivated by increasing international pressures relating to climate change and emission-cutting issues. As the world's largest carbon emitter, China's participation in the post-Kyoto Protocol climate talks, along with growing domestic demand for cleaner air and better environment, has forced the central government to pay more attention towards mitigation imperatives and low-carbon sustainable development. Nevertheless, the pro-environment element in the central renewable energy strategy has not been able to permeate the various local levels of governance, where pro-growth ideology has long been entrenched and still dominates, and renewables are facing strong resistance from interest groups supporting coal-fired electricity or controversial low-carbon alternatives, with societal forces including environmental non-governmental organizations (NGOs) too weak to intervene (Figure 9.1).

Importantly, even at the central level, the Chinese government's prioritization of renewable energy industry has not been purely fueled by mitigation imperatives; instead, such a strategy to a large extent was motivated by the country's mercantile mentality focusing on export revenues, domestic employment, economic growth, and energy self-sufficiency. The outbreak of the global financial crisis in 2008 reinforced such mercantile elements in China's policy-making, when China's export-oriented manufacturing was hit hard by sluggish external demand, and its surging energy demand was encountering volatile oil and coal prices in the international markets. In the aftermath of the financial crisis, China prioritized policy supports towards renewable equipment manufacturing aimed at foreign demand instead of domestic consumption, which led to overexpansion of wind turbine and solar panel production and trade disputes with the European Union and United States. Faced with heightened tariffs from foreign countries, China had to make a substantive policy shift from a stage in which most subsidies were designed to influence factor conditions

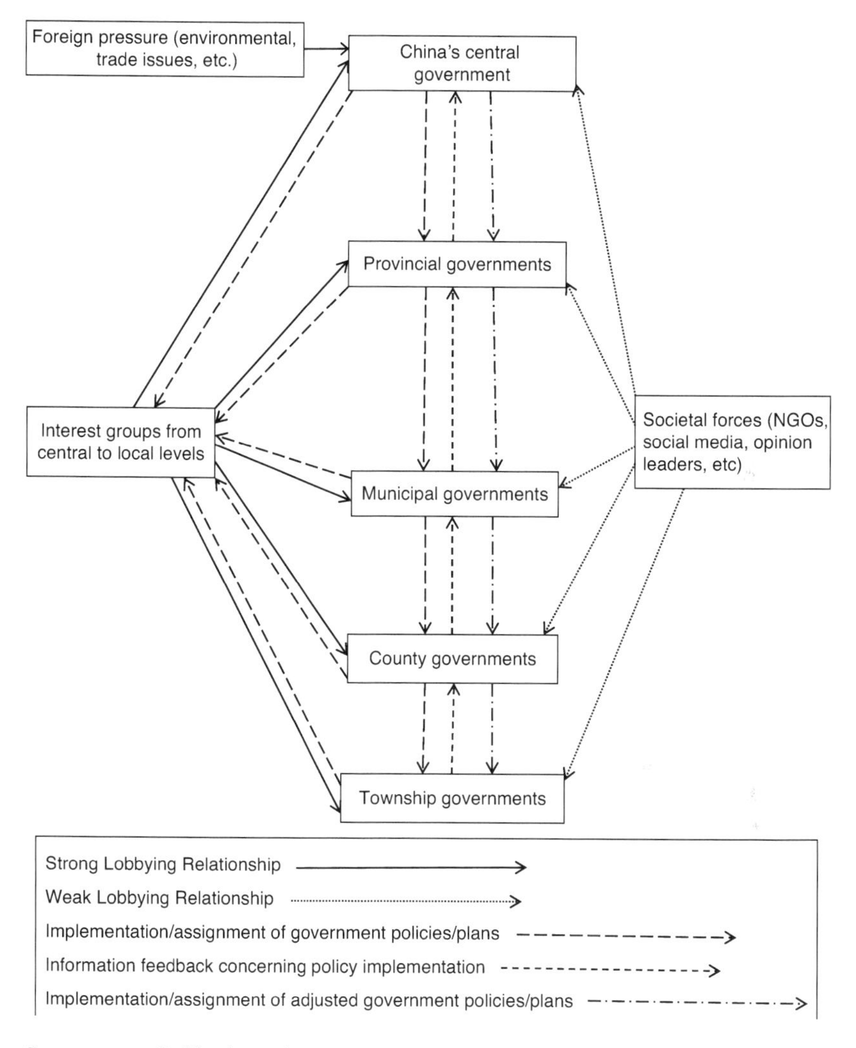

Source: compiled by the author

Figure 9.1 Policy interactions between Chinese central and local governments, foreign pressures, interest groups, and societal forces

and supporting industries, to a new stage of domestic demand stimulation with more subsidies aimed at supporting low-carbon power generation to absorb redundant manufacture capacity.

The shift from export-oriented equipment manufacturing toward domestic power generation led to a surge in newly added renewable capacity in both wind and solar power generation in both the 12th and the 13th Five-Year Plans (FYPs) periods, pushing China to the position of top renewable energy producer in the world. Nevertheless, due to imbalance of geographic distribution of renewable energy resources, as well as industrial hurdles such as power grid bottleneck, China's renewable energy ambition has resulted in huge waste and overcapacity in the northwestern and northern provinces, with the world's worst curtailment rates amidst excessive investment powered by massive government subsidies.

This study interestingly finds that China's centralized and top-down approach has been more effective in mobilizing local actors to achieve mercantile targets than in coercing them to fulfill pro-environment goals. The explanation of this phenomenon was made possible by the analysis of the lobbying relationships between various stakeholders (international community, domestic interest groups and societal forces) and the Chinese governments at various levels (see Figure 9.1). Compared with foreign governments and the international community, which can impose strong pressure mainly on the Chinese central government on environmental or trade issues, domestic energy-related interest groups are able to exert strong influences on both central and local governments at various levels.

Local actors, including government departments and corporations, are insensitive to international pressures and often spur the central leadership to enact pro-environment policies. In other words, international pressures, despite their strong impact on central leadership, usually fail to permeate down the various layers of China's local governance. As interest groups at various levels benefit more from mercantile policies, they tend to lobby local officials to implement pro-growth policies more forcefully while ignoring pro-environment policies. More often than not, they even encourage local governments to renegotiate with upper-level governments through providing feedback in their favor. After receiving reports from localities on first-phase policy implementation, the central government, based on the new information, needs to adjust its policies or plans for the next stage of implementation across the nation. Throughout the process, societal forces, including NGOs, social media, and opinion leaders, are basically marginalized and can only launch limited lobbying activities at various levels.

THE TYRANNY OF DISTANCE: CHALLENGES FROM THE LOCAL

China's geographic mismatch between resource-full power supply in the less-developed hinterland, in terms of both fossil fuels and low-carbon energy, and voracious energy demand in the resource-less coastal areas, reveals the permanent governance challenges facing central planners in the capital of a vast continental state. Blainey (1966) in his landmark book, *The Tyranny of Distance: How Distance Shaped Australia's History*, examines how Australia's geographic remoteness, particularly from its colonizer Great Britain and other continents, has reshaped Australian history, identity, and economic trajectory. Similarly, the remoteness of many large-scale energy projects in China, including windmills in Inner Mongolia, solar power projects in Gansu, and hydropower stations in Sichuan, reflects the insufficient knowledge of Chinese central planners regarding indigenous conditions in faraway counties, as well as Beijing's obsession with energy self-sufficiency ideology, and other political goals such as poverty alleviation. More economically efficient approaches to meet the coastal areas' expanding energy appetite include an increase in imports of energy products from other countries, or, more importantly, development of indigenous renewable energy projects including off-shore windmills and solar photovoltaics (PV) projects in eastern China.

In China's model of state capitalism, where FYPs and supportive industrial policies are, in many cases, treated more seriously by local stakeholders than market signals such as prices and costs, local governments and corporations are always enthusiastic about developing mega projects that can maximize growth rate. Nevertheless, the record-breaking pace of economic growth in statistical terms has not necessarily led to corresponding improvement in living standards, poverty reduction, or economic efficiency (Park and Wang 2001; Chen and Ravallion 2007; Donaldson 2011). When numerous wind turbines were installed in China's Gobi Desert under the central government's stimulus or subsidy policies, many local developers or officials did not even pause to consider whether they were accessible for power grids.

In an increasingly decentralized yet still non-democratic political context, or Chinese-style de facto federalism (Zheng 2007, p. 41), provincial, municipal, county, and township governments were endowed with much larger political and fiscal powers in handling local socioeconomic affairs. In a centralized system of cadre promotion, local officials had to bear more accountable responsibilities that may add risks to their political careers. When local governments increased their power bases vis-à-vis the central government in the beginning of the decentralization process, they were more interested in gaining authority over local economic affairs than in taking responsibility for environmental and social affairs. Local leaders, including the party bosses and administrative chiefs, all treat economic growth as top priority and are more willing to put hefty fiscal resources into capital-intensive fields such as manufacturing and infrastructure instead of paying attention to the efficiency of these local projects.

With regards to the implementation of energy policies at local levels, it is also subject to the interests of local governments and the level of coordination between the local and central government. The central government is responsible for policy regulation while the local governments are in charge of policy implementation. Unlike in the past when local levels used to be subject to a high degree of control from central government, such control has been diminishing in the reform era, and local governments have become more independent politically and economically. As they have their own interests to pursue, and their own interest groups to take care of, they need tactically to strike a balance between developing local economy and serving as political agents for the central government. Thus local governments sometimes prioritize local economic development and chose non-compliance or selective compliance with regards to orders from central government (Zhao 2001, p. 6). As observed by Michal Meidan, Philip Andrews-Speed, and Ma Xin, different regions of the country may follow different paths due to a number of reasons. As local governments have different priorities, sometimes they may justifiably reinterpret the national policies and combine them with more pressing issues that are of greater importance to local development; sometimes they may actively or passively obstruct the implementation of central government when the policy is not in their interest as regards development; or sometimes it is simply because the local government lacks the resources to implement national policies (Meidan et al. 2009, p. 616).

A good example of policy ineffectiveness at local levels is the implementation of the electricity pricing policy for the purpose of energy conservation in different regions, which reflects the 'twists and turns' of policy implementation at provincial and lower level governments, including how local governments have tried to 'deviate from the central orders while the central revises the policy from time to time to ensure local compliance' (Chen 2011). In 2004, six high energy-consuming industries including aluminum, ferroalloy, calcium carbide, caustic soda, cement, and the steel industries were targeted for an experiment with price differentiation. These industries were grouped into four categories according to their level of capacity, energy and resource efficiency, and environmental standards: encouraged, permitted, restricted, and eliminated. Problems emerged in 2006, after two years of implementation. Local governments worried that the implementation of the differential electricity pricing policy (DEPP) would hinder the economic growth and were reluctant to enforce the DEPP on those targeted industries. Besides, the allocation of revenues generated by the implementation of DEPP was unclear. In order to offset the negative impact on local economic development, some provinces even adopted a new policy, which was the preferential electricity pricing policy (PEPP). The PEPP allowed some high energy-consuming industries listed as 'encouraged' and 'permitted' to enjoy special electricity prices and lower their manufacturing cost. This problem persisted for more than a decade, leading to insufficient revenue from DEPP and which should have been used since 2009 for subsidizing renewable energy generation.

This central–local tension is also exemplified in the implementation of the coal liquefaction policy in China, in which local governments in coal-rich areas have pushed large investments into the technology despite the central government's call to slow down the development of liquefaction (Fang and Victor 2011). As observed, a major issue in studying central–local relations in China is the increasing resistance and distortion of central government policies at local levels (Zhong 2003, p. 129). After the initiation of reform and the opening up policy in China, the central government delegated power to local governments and also shared profits with them so that they had a greater incentive to develop the local economy (Peng 2009, p. 12). The post-Mao reforms devolved considerable financial and administrative control to the provinces and cities, counties, and towns, and now the central government wants to take back control, but the localities are loath to relinquish their new-found money-making powers (Cheek 2007, p. 108).

The ineffectiveness of policy implementation at local levels should also be attributed to the complexity and disjointed system of institutional structures among all the agencies in different levels. In the vertical dimension, central government and local governments are the two main actors regulating the renewable energy sector and supporting industry. However, there are multiple lines of government agencies in the vertical dimension that have direct interest in regulating renewable energy industry, including the National Development and Reform Commission (NDRC) and its National Energy Administration (NEA), Ministry of Natural Resources, Ministry of Ecological Environment, Ministry of Industry and Information Technology, Ministry of Finance and Ministry of Housing and Urban-Rural Development at all different levels of government. Consequently, the efforts required to adhere to all the relevant regulations and engage with all kinds of agency is so great that some local officials and stakeholders simply choose not to comply or to partially implement them. Besides, it is too difficult for higher levels of government to oversee the implementation of energy policies at local levels. Due to the complexity of the institutional structures and laxity in effective auditing and monitoring mechanisms, policy cheating at local levels is common.

THE RISE OF INTEREST GROUPS

The number and type of interest groups involved in energy policy-making have expanded substantially, with most ministries at the national level, big business entities, media, the military, local governments, NGOs, and even individuals playing increasingly significant roles in the whole process. Facing a much more complicated external and internal context, Chinese leaders have had to take on many other responsibilities and rely on others to help plan and implement Chinese energy policies, which further weakens the leadership's preferences while magnifying institutional and pluralistic impacts in the whole process. With the emergence of the renewable sectors, China's energy market has become more and more fragmented, resulting in greater barriers for implementation of central policies.

As perceived by outsiders, the agenda of state-owned enterprises (SOEs) and their development should have been controlled and regulated by government as there is a perception that the 'hidden hand of the socialist state lurks behind many Chinese companies' (Mooney 2005). However, the reality suggests otherwise. The term 'state-owned' does not equate to 'state-controlled.' Since the 1980s, China has undergone a series of decentralization and deregulation processes, including the transformation of energy ministries into SOEs. Through these reforms, the central government has empowered these companies with great autonomy in market operation. Although it still has some control over the SOEs through appointment of some high-level managers, energy pricing setting, and project approval, SOEs do enjoy great independence and autonomy in their daily operation. The SOEs have also seized the opportunity to drive the government's agenda and seek support for company development through many ways, including the facilitation of a friendly political environment with energy-rich countries, the approval for new projects both in domestic and foreign countries, and greater financial support such as low-interest loans and subsidies, and even compensation. In fact, the giant companies have strengthened their dominance in the national economy through significant contribution of taxes.

SOEs are adept at playing with the 'policy paralysis' (Downs 2008, p. 42) caused by the conflicts in interests and ambiguities in policies within energy bureaucracies, and advancing their corporate interests, sometimes even at the expense of national ones. Besides, even when government agencies are assigned to supervise SOEs, they do not have sufficient power to control the behavior of SOEs. For example, although the State Assets Supervision and Administration Commission (SASAC) is the formal owner of SOEs, it was formed much later than many energy giants and enjoys limited power over them. Consequently, the SOEs either resist implementation of policies that are considered unfavorable to their own development, or they integrate the national interest into their own new projects. As most energy SOEs have entrenched interests in conventional energy development due to their historical ties with the fossil fuel and hydropower sectors, they face conflicts of interest when asked by central authorities to be involved in renewable energy development.

In the Chinese model of state capitalism, the symbiotic relationship between corporations, both state and privately owned, and government departments has given the former great power to influence government decision-making and policy implementation. Compared with the ineffective government institutions, manufacturing and energy companies tend to be more active in lobbying the NDRC and other government agencies to support their capacity expansion or to take action to address problems that could have a negative impact on their expansion. More and more Chinese companies are showing an interest in corporate social responsibility (CSR), which concerns the relationship not just with their clients, suppliers, and employees, but also with other groups, and with social, ecological needs, values, and goals of the society (United Nations 2000, p. 2). In reality, however, many Chinese companies' involvement in low-carbon-related CSR is motivated by a number of self-interest goals ranging from propaganda of brands and improvement of company reputation to networking with government apparatus and potential customers. In many ways, corporations view environmental CSR as a means to secure financial gains, an essential component of their own development or even sometimes as the key to future organizational success (Bendell 2000, pp. 8–9). China's opaque political and economic structures, dominated by one-party rule, prevent the strict implementation of environmental laws, policies, and regulations. In particular, the strong collusion between corporations and the local and central governments is a crucial obstacle to public participation and environmental governance by citizens and NGOs (Chiashi 2017, p. 162).

THE NDRC/NEA DILEMMA

The pursuit of economic growth has been the single most important policy goal of emerging Asian countries in the recent past (Asian Development

Bank and Asian Development Bank Institute 2013, p. 3). In China, the core government apparatus to achieve the goal of sustained economic growth is the NDRC. It has been considered as a 'super ministry' or 'mini state council' as it has a broad mandate that covers almost all major economic policy planning, formulation, and implementation. Among all the ministries and government bodies, the NDRC is the leading institution taking charge of the administrative work of China's energy-related matters. Thus, the development-centric ideology rather than energy efficiency or environmental consciousness has been disseminated into the country's energy policy-making.

It is suggested that only when a genuine and far-reaching reform is carried out in the energy sector, especially when a new Ministry of Energy is created with sufficient authority and human resources, could effective policy implementation be achieved (Meidan et al. 2009, p. 613). At the time of writing, the long-anticipated Ministry of Energy has yet to be re-established, and vested interests in the status quo have sufficient influence to thwart such development. Rather, China has only conducted periodic restructuring of energy bureaucracy and produced multiple ministries and agencies that lacked the authority, autonomy, resources, and tools to govern the energy sector. China has undergone seven rounds of major administrative reorganization, respectively in 1982, 1988, 1993, 1998, 2003, 2008, and 2018, which have resulted in the abolishment of the Ministry of Energy and decentralization and redistribution of energy governance among multiple ministries. However, the responsibilities and roles are not clearly defined and policy overlaps and conflicts in interests are common. The governmental reform in 2018 weakened the NDRC by transferring part of its power over climate change issues to the Ministry of Ecological Environment. Even within NDRC, its energy authority is dispersed at least among five departments – the Department of Pricing, the Department of Basic Industries, the Department of High-Tech Industry, the Department of Resource Conservation and Environmental Protection, and the NEA.

The NEA was created in 2008 as the successor to the Energy Bureau under the NDRC. It also absorbed other energy offices in the NDRC, including the Office of the National Energy Leading Group and the nuclear power administration of the Commission of Science, Technology, and Industry for National Defense. Compared with the Energy Bureau, the NEA was given a broader mandate and vice-ministerial level authority – a step higher than the bureau-level authority enjoyed by the Energy Bureau. Its responsibilities include: planning, formulating, and implementing energy policies; supporting institutional reform; administering energy sectors; promoting energy conservation; reviewing and approving energy projects; enforcing energy reserve plans and policies; and conducting international energy cooperation missions. With staff numbers set at 240 (National Energy Administration of China 2018), the NEA does not have sufficient authority or human resources to coordinate the interests of government ministries, state-owned energy companies, and other agencies. Besides, the NDRC still maintains control over key issues, including price setting, as it has to achieve broader objectives such as combating inflation. Therefore, the NEA's mandate is still limited by the NDRC and it does not have sufficient authority to carry out its assigned tasks.

The provincial Development and Reform Commissions (DRCs) are offices of the NDRC at the provincial level. They carry out the tasks of the NDRC at local levels. However, as the institutional mechanisms of personnel and budgetary resource allocations are concentrated at the provincial level, the provincial DRCs have a closer relationship with local governments than with the NDRC. Leadership relationships between local DRCs and local governments are more obvious than with the NDRC. As a result, local DRCs which are supposed to be the agents of the NDRC in monitoring local governments, may eventually serve the interests of the local governments (Cheng and Tsai 2009, p. 122). The NDRC has the authority to approve large-scale energy projects. To avoid such red tape, local governments began splitting up projects into smaller ones that fell below this threshold (Weng et al. 2008). Many governments instruct their DRCs to approve small projects that may not meet the requirements set by NDRC. For example, local DRCs avoid sending small, low-efficiency electric power projects to NDRC, knowing that the NDRC will reject the projects because they are small and inefficient. Such collusion between local authorities and DRCs has made NDRC's policy implementation at the local levels unsuccessful (Chen 2011).

SUFFICIENCY INSTEAD OF EFFICIENCY

This study of China's low-carbon energy development shows that the country is more capable of handling the issue of self-sufficiency than the efficiency issue concerning energy production, transmission, and consumption. China has been improving the efficiency benchmark that includes reducing carbon intensity through raising the proportion of low-carbon energy and lowering the proportion of fossil fuel in its energy mix. Such a sufficiency-instead-of-efficiency strategy could, in the long run, lead to overdependence on newly added capacities of low-carbon energies including hydropower, nuclear, wind, solar, and geothermal power, and overlooking of the real efficiency issues addressed in documents such as the PRC Law on Conserving Energy or FYPs. In practice, overdevelopment of low-carbon alternatives has resulted in fierce competitions among different renewables, grid transmission bottlenecks, geographic mismatch between supply and demand, and consequently, overcapacity and low efficiency in the non-fossil fuel sector.

China, a country with flawed governance systems and an impatient eagerness to reap benefits from the accelerated globalization process, has had to face up to simultaneous serious environmental pollution and low economic efficiency even as billions of dollars of trade revenue flowed into the country. Although the central government has repeatedly launched 'macro-control' policy measures since 2004, the industrial overheating problem could not be effectively alleviated while the bottleneck in natural resources plus environmental pollution became even more serious. In the long run, the Chinese government would do well to introduce more economic incentives and disincentives to promote low-carbon development and curb pollution, instead of relying too much on short-term administrative measures such as eliminating redundant capacity or mandatory power purchase. Since an economic slowdown in China has been exacerbating the issue of excess industrial capacity, more domestic equipment suppliers in the wind and solar sectors are expected to be pushed to the edge of collapse without robust governmental subsidization of domestic solar PV and wind power generation.

More seriously, the huge investment of about 2.5 trillion yuan (US$361 billion) into renewable power generation between 2016 and 2020 (Reuters 2017) may bring about huge waste and inefficiency if the renewable energy developers fail to take into account the industrial bottlenecks caused by uneven geographic distributions in renewable sources, grid constraints, and policy disruptions from other local interest groups. The weak external renewable equipment demand and severe internal overcapacity have forced China's energy planners to make more of an effort to develop domestic low-carbon power sectors. Therefore, it is likely that the central government in the next few years will take actions to gradually absorb redundant local power capacity by building more ultra-high voltage (UHV) transmission networks, incentivizing local grid companies to purchase power from renewable power plants, and installing more solar panels and wind turbines in coastal provinces. Yet against the backdrop of economic slowdown and power overcapacity in most provinces, the surge in renewable energy production projects could exacerbate the existing overcapacity issue facing the whole power industry in China, leading to even lower investment returns and higher curtailment rates.

Besides, energy policies in different fields lack coordination, making it difficult to form a long-term and effective system to support sustainable renewable energy development. For instance, in order to reduce the investment cost in wind energy power plants, China cancelled the tariff for importing wind energy power generators, while on the other side, the concerned sectors were making active efforts to support China-based manufacturing process of wind power generation equipment, and implementing some model projects for the China-based manufacturing process (Zhang et al. 2009, p. 444). The two policies targeted different goals of efficiency and self-sufficiency, causing confusion in policy implementation and thus undermining the effectiveness of the overall industrial policy. China's progress in promoting non-fossil fuels is a natural extension of the country's preferred 'no-regret' strategy that emphasizes mitigation actions providing fringe benefits like profitability and employment in the country, regardless of whether the threat of climate change is real. For China, such an approach would take as a given that economic growth remains the top priority (Hatch 2003, p. 55). When such renewable energy policies were designed, policy-makers may have been concerned more about the economic factors such as energy supply and jobs rather than mitigating climate change.

China possesses one of the world's most complicated energy planning systems, with legal and policy frameworks in place to sustain it. In reality, implementation of these policies is often met with partial success in the authoritarian party-state. The institutional constraints to effective implementation of policy initiatives are numerous, and include the vague and contradictory nature of the relevant laws and regulations; the nature of the economic incentives for local government officials to prioritize economic growth at the expense of energy efficiency and the environment; the weakness of formal legal mechanisms; the close relationship between business and government; and the deep-rooted belief across society of the need for and desirability of economic advancement (Meidan et al. 2009, p. 615). Other reasons include the safety of transportation and transmission from distant areas to centers of demand; and, on many occasions, the frustration in enforcing energy policies domestically poses great challenges for the efficient development of China's resourceful low-carbon energy reserves.

While China has been experiencing a great leap forward in developing wind and solar PV power generation, policy-makers and market participants have been encountering industrial and technological barriers as well as geographic challenges that prevent full utilization of the quickly expanding installed capacity from these renewable resources. Moreover, when competing with conventional coal-fired thermal power that is much more polluting, wind and PV power production usually gets less support and subsidies from local governments, which exacerbates the market distortion in favor of thermal power plants, one of the major emission sources of China's urban smog and PM 2.5 (particulate matter with diameter less than or equal to 2.5 μm). Each year a huge proportion of wind and PV capacity remains unused. The integration of wind and PV generation especially in the western provinces of China has lagged behind capacity expansion, resulting in appalling waste. The Chinese government has realized the urgency of addressing these barriers and problems, to adjust supportive policies, speed up grid construction, and encourage more renewable capacity to be installed in the economically vibrant coastal provinces rather than the hinterland. However, such an effort may take years to have effect on existing renewable power production, transmission, consumption, and geographic redistribution.

REFERENCES

Asian Development Bank (ADB) and Asian Development Bank Institute (2013), *Low-Carbon Green Growth in Asia: Policies and Practices*, Hong Kong: ADB publication.

Bendell, J. (2000), *Terms for Endearment: Business, NGOs and Sustainable Development*, Sheffield: Greenleaf Publishing.

Blainey, G. N. (1966), *The Tyranny of Distance: How Distance Shaped Australia's History*, Melbourne: Sun Books.

Cheek, T. (2007), *Living With Reform: China Since 1989*, New York: Zed Books.

Chen, J. (2011), 'China's experiment on the differential electricity pricing policy and the struggle for energy conservation', *Energy Policy*, 39 (9), 5076–85.

Chen, S. and M. Ravallion (2007), 'Absolute poverty measures for the developing world, 1981–2004', Policy Research Working Paper no. 4211, Washington, DC: World Bank.

Cheng, T.-J. and C.-M. Tsai (2009), 'Powering rent seeking in the electricity industry', in T. W. Ngo and Y. Wu (eds), *Rent Seeking in China*, New York: Routledge, pp. 117–44.

Chiashi, A. (2017), 'Multi-tiered nature of environmental pollution problems and the pollution control governance in China: the role of environmental NGOs', in H. Kitagawa (ed.), *Environmental Policy and Governance in China*, Japan: Springer Japan KK, pp. 159–76.

Donaldson, J. A. (2011), *Small Works: Poverty and Economic Development in Southwestern China,* Ithaca: Cornell University Press.

Downs, E. S. (2008), 'China's "new" energy administration', *China Business Review*, November–December, 42–5.

Fang, R. and D. Victor (2011), 'Coal liquefaction policy in China: explaining the policy reversal since 2006', *Energy Policy*, 39 (12), 8175–84.

Hatch, M. (2003), 'Chinese politics, energy policy, and the international climate change negotiations', in P. G. Harris (ed.), *Global Warming and East Asia: The Domestic and International Politics of Climate Change*, London: Routledge, pp. 43–65.

Kang, J. and J. Su (2016), 'On supply-side reform', *China Economist*, 11 (5), 16–27.

Li, L. and H. Quan (2017), 'The logic of long-term growth of China: from new normal to supply-side reform', *International Journal of China Studies*, 8 (2), 139–60.
Meidan, M., P. Andrews-Speed, and M. Xin (2009), 'Shaping China's energy policy: actors and progresses', *Journal of Contemporary China*, 18 (61), 591–616.
Mooney, P. (2005), 'Undue fears of China, Inc.?', *YaleGlobal Online*, accessed 2 March 2018 at https://yaleglobal.yale.edu/content/undue-fears-china-inc.
National Energy Administration of China (2018), 'Introduction of National Energy Administration' [*guojia nengyuanju jianjie*], accessed 2 March 2018 at http://www.nea.gov.cn/gjnyj/index.htm.
Park, A. and S. Wang (2001), 'China's poverty statistics', *China Economic Review* 12 (4), 384–98.
Peng, W. (2009), *The Evolution of China's Coal Institutions*, Working Papers Series No. 86, Freeman Spogli Institute for International Studies.
Reuters (2017), 'China to plow $361 billion into renewable fuel by 2020', 5 Jan 2017, accessed 9 May 2018 at https://www.reuters.com/article/us-china-energy-renewables/china-to-plow-361-billion-into-renewable-fuel-by-2020-idUSKBN14P06P.
United Nations (2000), *Development of Guidelines on the Role and Social Responsibilities of the Private Sector*, Report of the Secretary General to the Preparatory Committee for the special session of the UN General Assembly on the World Summit for Social Development and Beyond (A/AC.253/21), New York: United Nations.
Weng, S., Z. Yuan, and Cheng, Z. (2008), 'China's congress revitalizes NDRC', *Economic Observer*, 21 March 2008, accessed on 2 March 2016 at http://www.eeo.com.cn/ens/2008/0321/94671.shtml.
Zhang, P., Y. Yang, J. Shi, Y. Zheng, L. Wang, and X. Li (2009), 'Opportunities and challenges for renewable energy policy in China', *Renewable and Sustainable Energy Reviews*, 13 (2), 439–49.
Zhao, J. (2001), *Reform of China's Energy Institutions and Policies: Historical Evolution and Current Challenges*, BCSIA Discussion Paper 2001–20, Boston, MA: Energy Technology Innovation Project, Kennedy School of Government, Harvard University.
Zheng, Y. (2007), *De Facto Federalism in China: Reforms and Dynamics of Central-Local Relations*, Singapore: World Scientific.
Zhong, Y. (2003), *Local Government and Politics in China: Challenges from Below*. Armonk, NY: M.E. Sharpe.

Index